The Honey Bee Around & About

The Honey Bee
Around & About

Celia F Davis

Bee Craft Limited

© 2007, 2014 Celia F Davis

First published in 2007 by Bee Craft Ltd. This second edition published in 2014.

All rights reserved. No part of this publication may be reproduced, stored in a retrieval system or transmitted, in any form or by any means, electronic, mechanical, photocopying, recording or otherwise, without the prior written permission of the publisher.

A catalogue record for this book is available from the British Library.

ISBN: 978-0-900147-15-9

Published in Great Britain by

Bee Craft Limited
National Beekeeping Centre
Stoneleigh Park
Stoneleigh
Warwickshire
CV8 2LG

Typeset by Buzzwords Editorial Ltd, Little Addington, Kettering, Northamptonshire NN14 4AX.

Printed in Great Britain by
Cambrian Printers, Aberystwyth

ACKNOWLEDGEMENTS

The lower diagram on page 64 is redrawn from *Honey Bee Pathology*, L Bailey and BV Ball, Figure 18, page 69, 1991, with kind permission of Elsevier.
Photographs by Celia and Cyril Davis except for those on page 70 (Beulah Cullen), page 96 (Barnaby Chambers), pages 57(b), 75 and 91 and figure on page 97 (National Bee Unit), pages 25–26, 29(a), 36a, 46, 67, 92(b), 99(b) and 148 (Adrian Waring), pages 18–21, 36(b), 51(b), 62, 92(a), 131(a), 144, 151, 157 and 166(a) (Claire Waring).

CONTENTS

	PREFACE	vii
	FORWARD	ix
1	HONEY BEE BEGINNINGS	1
	Knowing their place	1
	Origins	6
	Solitary wasps	6
	Social wasps	7
	Solitary bees	8
	Bumblebees	9
	Bees from wasps	11
	Climbing the social ladder	12
2	VARIATIONS	15
	Variety is the spice of life	15
	General classification of subspecies	17
	Only the best	23
	Raising new queens	26
	More advanced bee breeding	28
	Conclusions	31
3	HEALTH AND HAZARDS	33
	Keeping them healthy	33
	Types of organism causing disease in honey bees	38
4	WHEN THE KIDS ARE ILL	45
	Viruses affecting brood	45
	Bacteria affecting brood	47
	Fungi affecting brood	55
5	PROBLEMS WITH THE GROWN-UPS	59
	Viruses causing adult bee diseases	59
	Disease caused by microsporidia	63
	A disease caused by a protozoan	68
	Diseases caused by mites	70
6	MIGHTY MITES	75
	The nature of the beast	75
	Build-up in the colony	78
	Effects of the mite	80

	Chemicals and varroa	81
	Non-chemical control methods	85
	Long-term solutions?	87
	Another potential problem	87
7	THE ENEMY WITHIN AND WITHOUT	89
	One of life's little puzzles	89
	Small hive beetles	91
	Wax moths	91
	Wasps	95
	The Asian hornet	96
	Mice	97
	Woodpeckers	98
	An ongoing story	99
	Summing up	99
8	THE BROADER PICTURE	101
	The current situation	101
	Agricultural chemicals	106
	Stress on bees	109
	Be positive	110
9	FLOWER SHOW	111
	How flowers are made	111
	How flowers work	113
	Variation in structure and function in different plants	118
10	FOOD OF THE GODS	123
	Nectar	123
	Extra-floral nectaries	130
	Recycling – bee style	131
11	OTHER BEE ESSENTIALS	135
	Small and perfectly formed – pollen	135
	Sticky stuff	141
	Water	144
	Beeswax	144
12	ON BEST BEHAVIOUR	151
13	MORE THAN SWEET, THICK AND STICKY	159
	Nectar to honey	160
	Honey behaviour	164
	Granulation	166
	Fermentation – friend and foe	168
	Legal requirements for the sale of honey	171
	GLOSSARY	175
	REFERENCES AND FURTHER READING	181
	INDEX	183

PREFACE TO THE FIRST EDITION

In this second book, based on some of the articles I have written for *Bee Craft* over the past 10 years, I have looked at the honey bee in a wider context. It is really a book in four parts: describing the place of our bees in the grand scheme of things, their relationship to some of the organisms that share their world and often make them sick, flowers and the close co-operations between them and their honey bee visitors and, finally, a section on honey, an extraordinary substance vital for the survival of both bees and beekeepers. As before, my aim has been to make the text readable and accessible to the non-scientist.

Writing is largely a solitary occupation but I have been much encouraged by the reaction to my first book and the enthusiasm of many of my readers has kept my nose firmly fixed to the grindstone when more attractive occupations threatened to distract me. I am grateful to them all.

Once the solitary part is over, a number of other individuals leap into action to convert the rough material into a finished book and here I must mention all those who have contributed to that process. First on the list are my editors, Bee Craft Ltd under the Chairmanship of Andrew Gibb. Their continued support has been invaluable. My sincerest thanks must be reserved for Claire Waring who has been my friend for many years and without whose encouragement I would probably never have started writing articles, let alone books. She has edited this book with her usual confidence and efficiency and the finished product is a tribute to her professionalism and skill. I am also grateful to Norman Carreck for his help and advice.

On the family front, my daughter, Sarah, read and corrected the early drafts while juggling her busy life of work, home, garden and beekeeping, and Cyril, my husband, often had to endure periods when bees and writing seemed to take precedence over other matters, but he still managed to remain supportive and encouraging.

PREFACE

The stars of this book must always remain the bees. Who could imagine a world without them? Certainly they are essential for the production of a great deal of food and are a vital part of the environment as a whole. In addition, they produce that most delectable of natural foods, honey, and enrich the lives of those of us fortunate enough to work with them.

Celia Davis, NDB
March 2007

PREFACE TO THE SECOND EDITION

It is now nearly seven years since the first edition of *The Honey Bee Around and About* was published and the beekeeping world has moved on considerably in that time. One of the exciting things about keeping bees is the opportunity, and need, to embrace new ideas and developments and I have tried to address this. This has been particularly so in the chapters on health and disease, subjects which, sadly, have assumed greater importance to us all in recent years. I have also written a new chapter which seeks to paint a broader picture of the challenges which are facing beekeeping and provide a more holistic approach to the problems.

Once again I have to thank my daughter and beekeeping partner, Sarah, for her helpful comments and Bee Craft Ltd for its continuing support. Special thanks go to Claire Waring for her help, advice and editorial knowledge and experience which have been invaluable and without which this book would not exist.

Celia F Davis, NDB
March 2014

FOREWORD TO THE FIRST EDITION

Is beekeeping an art, a craft, a science, or a combination of all three? Part of its attraction for me has always been that it covers so many different subjects, but there can be no doubt that successful beekeeping depends to a large extent on the application of scientific knowledge. One of the problems facing beekeepers, however, is that scientific information is often surprisingly difficult to obtain. Much scientific research is published in specialist journals which are not generally available to the public. Although most journals now have articles available online, the costs of downloading even one paper can be extortionate. Membership of the International Bee Research Association provides access to papers, but they often appear written in an arcane style not readily accessible to the average beekeeper.

This is where Celia's first book, *The Honey Bee Inside Out*, came into its own, but condensing a huge amount of technical literature into a readily accessible form, thoroughly expressing often very complex ideas in a clear style. That book covered the anatomy of the honey bee and its behaviour, with the help of clear diagrams and beautiful photographs. It rapidly joined the recommended reading list for the BBKA examinations and, following its very good reception, she has now produced this volume, which covers most of the remaining important topics in bee biology.

The first part of the book considers the place of bees in the natural world, particularly in relation to their close relatives, the wasps and other bee species. Solitary bees and bumblebees are usually neglected in beekeeping books, yet many beekeepers are now becoming increasingly aware of the need to understand, and to conserve, these important and often threatened pollinators, which depend on many of the same resources as our managed honey bees.

The evolution of sociality (on several separate occasions) in the insects is a fascinating topic on its own, but of course many of the management decisions that we make in beekeeping are entirely dependant on a good understanding of honey bee social behaviour. Equally important is an understanding of the characteristics of the

FORWARD

different subspecies of honey bees found and used commercially around the world, a knowledge of the benefits of careful queen rearing and bee breeding, and of the special problems inherent in breeding bees compared with other farm animals.

A large part of the book them considers the various pests and pathogens that cause the diseases of honey bees. This is very timely, as many beekeepers both here in Britain, but also in continental Europe and North America, are currently experiencing the serious problems that have occurred due to the varroa mite having developed resistance to the most widely used chemical controls. This problem makes the understanding of the biology of the mite and its associated pathogens even more important. It is only with an intimate knowledge of the nature of the problem that we can devise new strategies to deal with it.

In the first part of the book the intimate relationship between bees and flowers and the coevolution was explained. Now Celia turns to the flowers themselves, explaining how they work and how different designs provide pollen and nectar in various ways, the process of nectar secretion and the factors which influence it, extra-floral nectaries and the production of honeydew, and the adaptations of bees to carry pollen and nectar.

Finally, Celia turns to bee foraging behaviour and the hive products that result, including propolis and beeswax and honey itself. She also covers the often tortuous legal requirements for preparing it for sale.

Once again, this book will prove immensely valuable to those taking exams, but should be on the bookshelf of every beekeeper, especially those intending to teach others.

Norman J Carreck, NDB
International Bee Research Association, Cardiff
March 2007

1 HONEY BEE BEGINNINGS

KNOWING THEIR PLACE

Where do honey bees belong? Where, in the great scheme of things, do we put this small, brown, but highly evolved and complex creature which we find so infinitely fascinating?

To understand its place we must look at the way living things are related to one another. This basic system of classifying all living organisms was devised by Linnaeus (1707–88) so it is very old. The largest groups in the system contain many different types of organisms but these groups are divided into smaller ones, and so on. As we progress we find that each sub-group contains fewer organisms but with greater similarity to one another and the final sub-group includes just one type of individual. Each group of organisms is called a *taxon* and the study of the classification of organisms into taxons is called taxonomy.

It may seem a very dry and uninteresting aspect of biology but it is fraught with argument and is changing all the time as new discoveries are made. Linnaeus gave each taxon a name and these constitute the major Linnaean groups, shown in the example below in UPPER CASE, however there are 'extra' groups slotted in between the major ones and these are shown in lower case. These non-Linnaean groups are:

- sub-groups if they occur after the main group and contain fewer organisms
- super-groups, if they occur before the main group and contain more organisms.

Of course, the honey bees are completely oblivious to their place in the biologists' schemes, but classification does enable us to understand any organism better, to see its relationships with other organisms and to understand how it has evolved to its present place

CHAPTER 1

Apis mellifera, small and brown but highly evolved and complex

in the world. An example, using our beloved honey bee, will enable us to see how classification works and the honey bee's place in it.

1. KINGDOM: **Animal**
   This is a huge group and includes all those organisms in the world which are animals. So this is the first group to which our bee belongs. (There are four other Kingdoms and between them they contain all the living organisms on the planet.)
   Another example is the Fungi Kingdom.

2. PHYLUM: **Arthropoda**
   The Animal Kingdom is divided into a number of Phyla and this is the biggest one. It includes a host of weird and wonderful creatures. They all have a body divided into sections, called segments, enclosed in a hard outer case, and some of the segments carry appendages such as legs or wings. There is a head carrying sense-organs and mouthparts. The word 'Arthropoda' comes from two Greek words, *arthron*, a joint, and *podos*, a foot, and is a bit of a misnomer as the appendages, and not just the feet, are jointed. The Arthropods include creatures such as centipedes, lobsters, crabs, spiders and mites, as well as many other, sometimes obscure, types and, of course, the honey bee.

A harvestman
A member of the phylum Arthropoda

HONEY BEE BEGINNINGS

Other examples of Phyla are Mollusca (snails, squids, bivalve shellfish, etc.) and Annelida (segmented worms including the earthworms and leeches).

3 CLASS: **Insecta**
Now we are getting onto familiar territory. All insects have a head, thorax and abdomen. On the head they carry the mouthparts and sensory organs, the thorax has three segments, each carrying one pair of legs. So far the insect seems to have a great affinity for the number three but, when it comes to wings they have a pair of wings on each of the last two segments, or a pair on the middle thoracic segment only, or none at all. The abdomen does not have any legs. Examples of other Classes within the Arthropoda are Arachnida (including spiders, mites, ticks, scorpions, harvestmen and many other similar types) and Diplopoda (millipedes, having two pairs of legs on most segments).

4 ORDER: **Hymenoptera**
Insects with two pairs of membranous wings which can be linked together. Twenty-nine orders of insects are usually recognised, including the Lepidoptera (butterflies and moths) and the Diptera (flies).

5 Sub-order: **Apocrita**
This is the first 'extra' group we have included. Because it comes after (or below) an Order it is called a sub-order and is a section of the Hymenoptera. All the Apocrita have a 'wasp waist' (petiole) which is a very narrow extension of the second abdominal segment A2, structurally forming a flexible connection between the thorax and the abdomen of the insect. It follows then that the first abdominal segment, called the propodaeum, is attached

Typically insects have three parts to their bodies and three pairs of legs on the thorax

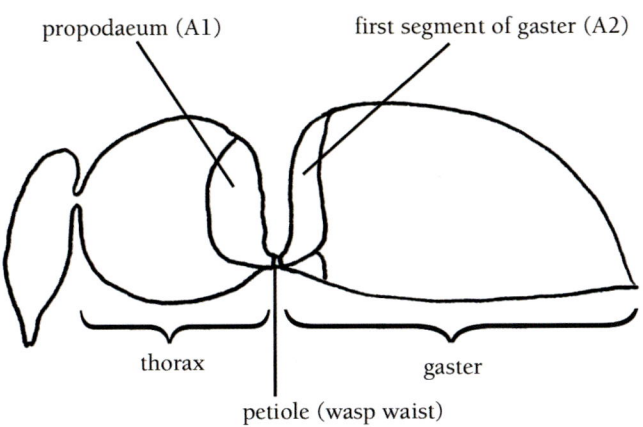

Features of the sub-order Apocrita

CHAPTER 1

The larvae of *Cotesia glomerata* leave a butterfly larva (*Pieris brassicae*) which they have parasitised

Other insects belonging to the Parasitica induce growth such as this bedeguar gall on rose

to the thorax. The rest of the abdomen, behind the waist, is now called the gaster. The larvae of the Apocrita are simple legless grubs because they all live in very protected environments where food is provided for them one way or another. There is only one other sub-order of the Hymenoptera. It is called the Symphyta and includes the sawflies and woodwasps. These are quite stout insects without a wasp waist and their larvae are active, with legs.

6 Division: **Aculeata**

This is another non-Linnaean group but we are getting really close now because the Aculeata includes the bees, ants and hunting wasps. They are all characterised by having a modified ovipositor. In other insects this is the structure down which an egg passes as it is laid, but in the Aculeata this has now become a sting and the eggs no longer pass down it. (The Latin word *aculeas* means sword.)

The other Division of the Apocrita is the Parasitica, including all the gall wasps, which induce so many extraordinary galls on plants, and an enormous number of little 'wasps' which are all parasitic on other insects. If you really want something to study, there is very little known about most of these and you could make your mark, providing you have good eyesight, a lot of patience and you recognise that your fan club will be small.

7 Superfamily: **Apoidea**

This group appears before (or above) the Linnaean group 'Family' so it is given the prefix 'super'. All the bees are in this group. They are set apart from the wasps because they feed on the products of plants and are not carnivorous, either in the adult or larval stage, so the basic difference between wasps and bees is really one of behaviour. As a result of this behaviour, bees have developed structures for transporting pollen and their hairs have become branched, usually described as 'plumose'. Within the Division Aculeata there are several other superfamilies, one of which is the Formicoidea (ants).

8 FAMILY: **Apidae**

We have reached the group which includes the honey bees and other highly social bees. A familiar example of another family is the Megachilidae (leaf-cutting bees). These are solitary bees which line their nests with little pieces of leaf which they usually cut, very neatly, from roses, to the irritation of some rose-growers.

9 GENUS: ***Apis***

The true honey bees. This generic name is used as the first name of the organism concerned, in this case the honey bee.

Bombus is another genus in the Apidae.

HONEY BEE BEGINNINGS

10 SPECIES: *mellifera*
   The western honey bee. This is the second, specific, name for our organism. Another species is *cerana*.

So, the Western honey bee, the creature we keep in our hives, is called **Apis mellifera**. Notice that each species has two names. The first name, which is the genus to which the organism belongs, always starts with a capital letter while the second, which may be called the 'specific' name, because it denotes the species, always starts with a lower case letter. They are always written either *in italics* or <u>underlined</u>. When the genus is known, it is often just given an initial, eg, *A. mellifera*.

I think it may help to summarise the classification of the honey bee given above in a simple table which is shown below. The other examples given bear no relationship to one another.

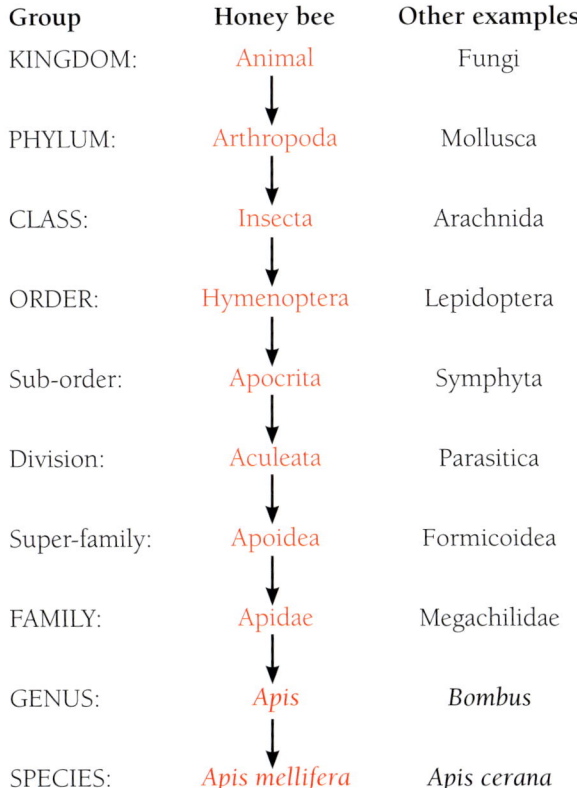

Group	Honey bee	Other examples
KINGDOM:	Animal	Fungi
PHYLUM:	Arthropoda	Mollusca
CLASS:	Insecta	Arachnida
ORDER:	Hymenoptera	Lepidoptera
Sub-order:	Apocrita	Symphyta
Division:	Aculeata	Parasitica
Super-family:	Apoidea	Formicoidea
FAMILY:	Apidae	Megachilidae
GENUS:	*Apis*	*Bombus*
SPECIES:	*Apis mellifera*	*Apis cerana*

Remember that within this classification it is possible to slot in other groups, but I have tried to keep it simple without losing the important points.

CHAPTER 1

Mellinus arvensis, a solitary wasp, on Golden Rod (*Solidago* sp.). This wasp feeds its larvae on flies

The nest of *M. arvensis* is made in the ground

Ammophila sabulosa, a solitary wasp, shown at its nest with its prey, the larva of the Pine Beauty Moth (*Panolis flammea*)

ORIGINS

Where did we come from? A familiar question and we are probably all fairly au fait with our own development from some kind of ape ancestor, but what about our bees? How, and when, did bees develop into these very complex communities that we keep and try to control today? Before we go on to answer that question we are going to look at the lives of a few of their relations. All these are members of the Hymenoptera but none of them is as highly evolved as the honey bee. All comments apply to insects in the UK only.

SOLITARY WASPS

There are about 200 species of these in the UK, but their life cycles are all similar. So where do you look for them? They make their nests in a variety of places:

- holes in wood where they may take over the tunnels of beetles
- broken stems of plants
- holes in mortar
- holes in soil, often in sandy areas and on banks.

They frequently nest in great numbers although they are completely solitary and, because the adult wasps feed on nectar, they can often be found on flowers.

The larvae are carnivorous (with the exception of one group) and the method of feeding could easily be used as a basis for a horror movie. Each species of wasp has its own particular type of prey, some being so specific that they concentrate on one or two species only.

The list is long but, as examples, we can mention the larvae of butterflies and moths, flies, weevils, aphids and spiders. Because the female wasp provisions her nest and leaves the larvae to feed themselves (although some species do bring in prey over a period of time), she has to ensure that the 'meat' keeps fresh so she does not kill it, but merely paralyses it, usually by stinging. As a special twist to this saga, some prey items, especially caterpillars, have been shown to have the larvae of other parasites still developing within them inside the wasp's nest. Female solitary wasps are intrepid hunters, seeking out and immobilising prey which, in many cases, is bigger than the wasp itself. Once helpless, the luckless prey is dragged or carried back to the wasp's nest. When there is sufficient food, the female wasp lays an egg on it and then seals the entrance to the nest hole. The wasp larva, on hatching, devours the (still living) food and eventually develops into an adult wasp which emerges to mate and start the cycle all over again.

There are many variations on this general pattern, some wasps even specialising in stealing food from other solitary wasps' nests. The adult wasps are very active insects, as you would expect from their lifestyle. They may have the yellow and black colouration we associate with wasps or they may have other markings, in the orange/red range, or be completely black. The species in the photographs illustrate this variation in colouring. *Ammophila sabulosa* is sometimes referred to as a sand wasp because it makes its nest in sandy soil and, where there is suitable habitat, it is often possible to find an enormous number of these wasps all living in the same area. It is a caterpillar specialist, stocking its nest with one single, large larva and is also one of the largest solitary wasps, measuring around 20 mm in length, excluding its antennae, so it is easy to spot. *Mellinus arvensis* catches flies and also nests in the soil where it constructs little entrance porches of soil over its nest.

A queen of the common wasp (*Vespula vulgaris*) in late summer. She will soon begin hibernation

Worker of the common wasp (*V. vulgaris*)
(*Note the face pattern which is used in identification*)

SOCIAL WASPS

Social wasps are much more familiar insects and their life cycles are better known, but many people mistake them for bees. There are eight British species plus one cuckoo wasp and they are all superficially similar as far as colouring goes, although the hornet, the largest of the group, is reddish brown rather than black.

The wasp nest is a work of art constructed from 'paper,' made from wood scraped from worn or decaying timber and mixed with saliva. There are three wasp genera (plural of genus):

- *Vespula*. This genus includes the two common wasps, *V. vulgaris* and *V. germanica*. They choose sheltered sites for their nests such as holes in the ground, lofts, bird nest boxes, and so on and are very common in the summer, causing annoyance to many people. There are two more species in this genus, *V. rufa* and *V. austriaca*. The latter is a wasp living as a cuckoo in the nests of *V. rufa*.
- *Dolichovespula*. These wasps nest in the open, suspending the structure from twigs and branches in a tree or bush. There are four species altogether in this genus.
- *Vespa*. This genus contains only one species, the European hornet, *V. crabro*. It generally builds its nest in holes in trees but does sometimes use buildings such as barns. Because it

The nest of the median wasp (*Dolichovespula media*) is built in the open in a bush or tree. The pattern is characteristic

V. crabro, the hornet, is our largest wasp

is so big, the hornet is often a source of fear but it is truly a gentle giant unless something attacks its nest.

The larvae of all the social wasps, like those of the solitary wasps, are carnivorous but, just like the solitary wasps, the adults eat sweet things such as nectar and derive much of their food from a secretion produced in the salivary glands of the larvae.

The annual cycle can be summarised as follows:

- queens which have hibernated over the winter emerge in spring, feed on flowers and choose nest sites.
- the queen builds a small nest, lays her first eggs and catches prey to feed the larvae.
- the first worker wasps emerge and take over the work of brood rearing and nest building.
- in summer, when the nest has expanded to a good size, male and female (queen) wasps are produced. The cue for this may be the slightly shortening days after mid-summer.
- males and females mate.
- in late summer, after the reproductive forms have been produced, the colony begins to lose its cohesion and wasps become a nuisance because they have no larvae to feed and therefore no sweet saliva to eat. They go in search of sweet foods such as nectar, ripe fruits, jam and honey.
- when the weather gets colder, nests die out and young, mated queens hibernate. All the other wasps die, although the males may eke out an existence on late flowers such as ivy until the first frosts come along. They are often a pitiful sight on cold mornings in late autumn.
- the nest structure disintegrates. Wasp nests are very fragile structures unless they are well-maintained or well-protected.

The social wasps have developed a community, but that community has to begin each season all over again with a new founding 'mother' and, during the early stages of development, that mother has to do everything, including nest-building, hunting and caring for her offspring.

SOLITARY BEES

There are around 250 species of solitary bees, belonging to six different Families, in the UK, and their life cycles are very similar to those of the solitary wasps which we have already discussed. The major difference is that, because they are bees, their larvae, as well

The female tawny mining bee (*Andrena fulva*) collecting pollen in the spring

as the adults, feed exclusively on the products of flowers, not on meat, so they provision their nests with pollen, which the female collects and carries back to the nest. To enable the females to do this efficiently, bees have developed branched (plumose) hairs, which are characteristic of all bees, whether they are females or males, and even those that are parasitic and never collect pollen. The pollen is usually carried in *scopae*, which are structures resembling brushes found either on the legs or beneath the abdomen, but one group transports it in the crop since it has no specialised pollen-carrying structures.

The adult bees use nectar from the flowers as an energy source and may use a little nectar to moisten the pollen as they transport it back to the nest, but they do not store nectar or make honey. The nests are found in similar types of habitat to those of the solitary wasps and you are most likely to come across the various mining bees. Some of these may cause problems when they choose, as their nest site, someone's carefully tended lawn. Like some of the solitary wasps, they tend to be gregarious and many hundreds of nests can occur together in 'villages'. This occasionally leads to 'swarm' calls from members of the public when the young bees all emerge at the same time. People are much more aware of solitary bees since the advent of solitary bee nest tubes, now generally available in most garden centres, and, in many cases, these bees are encouraged into gardens, providing interest and pollination services.

A pair of red mason bees (*Osmia rufa*) mating

Larvae of *O. rufa*
Each is in an individual cell with its own supply of pollen. This nest is inside a plant stem but O. Rufa is an opportunist nester and is usually the commonest bee in artificial bee tubes

BUMBLEBEES

We keep one species of bee in our hives but, as we have seen, there are many other species out there in the wild. These are not honey bees. Most of them are solitary bees, which we have already discussed, and 23 are bumblebees which belong to the family Apidae. Within that family all bumblebees belong to the genus *Bombus*. Honey bees are social insects, living in a large colony with a single egg-laying queen and many thousands of non-reproductive female workers which are her offspring. We are familiar with this insect and its behaviour. The true bumblebees are also social with the same basic nest organisation but on a smaller scale and with many significant differences.

The nests are annual affairs started by a single over-wintered queen in the spring and building up, even at the height of summer and in the most prolific species, to no more than a few hundred individuals. Their nest architecture lacks the uniformity and neatness of the honey bee. Nests may be constructed (depending on the species) in old mouse holes, in tussocky grass and in many

A *Bombus pascuorum* worker

CHAPTER 1

A *Bombus pascuorum* drone
Unlike honey bee drones, bumblebee males feed from flowers

other places where shelter is afforded. Beneath house floors, with access via an air brick, and bird nest boxes seem to be favourites and I have seen them in the bottom of an old chair and in a roll of carpet underfelt, both in out-houses where the door did not quite fit – very comfy accommodation indeed. The life cycles of all species follow a similar pattern although there are differences in detail:

- overwintered queens emerge from hibernation in March or April.
- queens feed on nectar and pollen and search for a suitable nest site.
- the queen provisions her nest with nectar and pollen and lays her first eggs.
- the queen incubates her eggs, sitting on them and feeding from her nectar store, until they hatch.
- the queen feeds her brood, occasionally leaving the nest to forage.
- the newly hatched workers take over the work of brooding, feeding brood and foraging, leaving the queen free to lay eggs.
- when the nest is well developed males and females are produced.
- males and females mate (I have observed this only twice, in each case near the ground).
- the nest breaks down and dies out, in some species as early as mid-summer although, like the wasps, the males can live for quite a long time and can often be found huddled under flowers on cold mornings. The nests are usually destroyed very quickly by wax moths.
- the mated queens carry on feeding to build up their fat bodies and then find a hibernation site.

Everyone seems to love bumblebees and even the most unobservant notice the huge queens when they emerge in the early spring. At this stage life is very difficult for a bumblebee queen and many nests fail for a variety of reasons.

No account of bumblebees would be complete without mention of the cuckoo bumblebees. These used to be in a separate genus, *Psithyrus*, but are now all included in *Bombus*. There are six species and they act just like the cuckoo bird, emerging after the social bumblebees have established their nests and taking them over. Each cuckoo is parasitic on one or two host species and, because they have no worker caste, they use the workers of their host to rear their young. The arrival of a cuckoo spells disaster for the original colony and this is another cause of great loss.

HONEY BEE BEGINNINGS

New arrivals

The number of species of wasps and bees present in the UK is not static. Some species become very rare and, eventually, extinct, while others are sometimes reintroduced or arrive as immigrants. One such is *Bombus hypnorum*, the Tree Bumblebee, which was first recorded in the UK in 2001 and has now spread to many parts and, in some areas, is quite common. Another, which at the time of writing is not a UK resident, is the Asian hornet, *Vespa velutina*. This may prove to be a problem for beekeepers here as it is a fearsome predator of other insects and can target hives, resulting in the loss of colonies. It is smaller than our native hornet and is much darker in colour (see page 96).

BEES FROM WASPS

I am now going to return to the question I asked some time ago: how, and when did honey bees evolve into the complex communities with which we are so familiar today? There is one certainty – they were around long before our very early ancestors descended from the trees and began to grunt to one another about the design of wooden clubs. That happened very recently – probably between one and two million years ago, but the first insects appeared around 400 million years ago, during the Silurian Period. During the next 100 million years, insects developed wings and there were recognisable insects such as giant dragonflies soaring around the huge forests which produced the coal measures, about 300 million years ago. There are various landmarks along the way and all the dates I give are approximations because different sources give different times and I was not around to check them! But then again, neither was anyone else and, in any case, a few million years is neither here nor there in the history of life.

The first bees evolved from solitary wasps belonging to the superfamily Sphecoidea. These were hunting wasps and would have been ancestors of the hunting wasps alive today. *Ammophila sabulosa* is a member of this superfamily. Wasp larvae are carnivorous but adult wasps like sweet food so it is likely that at some stage a wasp which had become accustomed to visiting flowers to sip the nectar, or one which had become used to capturing aphids containing sweet plant sap, began to feed its larvae on nectar and pollen instead of 'meat'. After all, it is a lot easier to visit flowers than chase around after prey which can run or fly away. This change probably happened around 140 million years ago, at the same time as the flowering plants (Angiospermae) became the dominant group of

Bees are all covered with branched (plumose) hairs. This sets them apart from the wasps

plants. This group and the bees have developed alongside each other and have become inter-dependent. Also at this time, some bees became social insects.

The first bees in the genus *Apis* probably appeared in the area which is now India, or the Asian countries just to the east of it, as this is where the largest number of bee species is still found, but remember that the world did not look as it does today. Between 300 and 200 million years ago there was one large land mass on Earth, which then separated into two. These subsequently split, and *A. mellifera* would have spread to the UK before it was separated from the rest of Europe. So, climate was very different and the early spread of species was unhindered by huge oceans. Although the early fossil record is not very good and few remnants of bees have been discovered, a bee around 70 million years old has been found preserved in amber. This was a species of ***Trigona*** and closely resembles other species still alive. More recently, a species with some wasp characteristics, but undoubtedly a small bee with plumose hairs, has been found in Burma and dated at 100 million years old, so this specimen appears to support the theories of wasp to bee evolution.

CLIMBING THE SOCIAL LADDER

The first bees would have been solitary, as are most bees today. It is safe to say that no bee ever woke up in the morning and said to other bees who happened to be around, 'Hey girls! Why don't we get together?' The development of bee communities was a gradual process. Although we were not around to see it, we can see all the stages that may have contributed to it in bee species alive today and we can recognise various significant steps.

1   Communal living where a number of female solitary bees use one entrance but each produces her own nest within the complex. These are all the same generation and may, or may not, be sisters. Each bee collects all the food she needs for her cell(s) and lays her own eggs, independent of all the others in the community. We see this situation each spring at home, where the Red Mason Bee, *Osmia rufa*, a normally solitary bee, lives communally behind one of our window sills. All the females use the same hole, but unfortunately it is impossible to see what happens once they get inside. The number of bees seems to be increasing as each year more of the bees hatching

from the nests return there, so it is clearly successful. This kind of communal living is not social but the individuals probably gain some benefit in deterring predators, because there is always someone at home, whereas the completely solitary bee has to leave its partly provisioned nest unattended for long periods of time while it is away collecting food for its offspring.

2. Simple levels of true sociality. These can take various forms:
   - the females all contribute to building the cells and collecting pollen for them but some lay eggs and others function as workers. Some may be unmated and so be non-reproductive. All the females present belong to the same generation but they may not be related to one another. There is no care for the offspring other than providing food and the resultant bees never see their mother(s). These may be called *semisocial* colonies and there are examples among some tropical bees.
   - a female, instead of just laying her eggs with a supply of food, tends her young by continuing to feed and protect them as they grow. The term we can use to describe this set-up is ***sub-social*** and in this situation we have a family group with a single mother at its head. In most cases, the mother dies before her young mature, but it is only a short step to those larvae becoming adults before mum dies. The potential for them helping mother rear the next generation is then in place. This is the first example of more than one generation co-operating and really gets us onto the road to being truly social.

The bee *Halictus rubicundus* is a very good example. The original female produces a brood of daughters who do not mate themselves but help their mother to rear a second brood of males and females. At this stage the nest is sub-social. The second generation of females mate and then hibernate to begin the cycle again the next year. However, the interesting thing about this little bee is that it can also function as a purely solitary bee, producing only one generation of males and females in the year. It behaves like this where the climate is not good enough to enable it to produce two generations, such as in the northern UK.

It may be worth mentioning here that sociality can go 'backwards' and, it is estimated that during the Cretaceous period, when bees were developing rapidly, a number of primitively social bees switched back to being solitary. But we are getting off the straight and narrow.

We now have to introduce a new term: ***Eusocial***. Eusocial bees consist of an egg-laying queen and adult worker females from one

Halictus rubicundus **shows the early stages of social development**

CHAPTER 1

A *Bombus pascuorum* nest
Bumblebee nests last for a few months only and never reach a large size

or more generations which are her offspring. We recognise *Apis mellifera* from this description, but other groups of bees also fit into this category.

The bumblebees, *Bombus* spp. are said to be *primitively eusocial* because, although at the height of summer their colonies appear to be the same as honey bee colonies, they start their colonies from scratch each year, with a single mated female. This means that they pass through a solitary stage before they become social. Compare this with the honey bee colony which is perennial and always exists as a huge family of several generations containing many individuals. This is an important point illustrating that individual species can pass through various different categories in the course of a life cycle.

The development of sociality in the Hymenoptera is an interesting study and we have only touched it very briefly. It certainly gives new meaning to the term 'climbing the social ladder' and, before we leave the topic completely, I want to mention termites (order: Isoptera). These are distant relatives of cockroaches and developed from them. Cockroaches are found in the order Dictyoptera and are very primitive insects which were around long before the first bees arrived on the scene. However, the social organisation of the termites and the social Hymenoptera shows many similarities such as use of pheromones and food sharing (trophallaxis), as well as many differences. In many ways these two quite unrelated groups have arrived at much the same point in social evolution and have probably reached their particular pinnacles of development because of the limitations of the insect brain. Insects also have a problem with size because their exoskeleton is bulky and constraining, so maybe we will never see the emergence of large, super termites or bees with more complicated societies and superior brains – but then who would have thought an ape would, one day, produce a computer or travel to the moon?

2 VARIATIONS

VARIETY IS THE SPICE OF LIFE

Too much uniformity is boring, so the fact that our honey bee, *Apis mellifera*, has many different subspecies (or races) adds interest and variety to life. Of course, the various forms of *A. mellifera* did not evolve just to provide us with extra interest, but in response to different conditions. We have seen that the original ancestors of the genus *Apis*, to which our honey bees belong, evolved many millions of years ago in India or nearby, but the first *Apis mellifera* (Western honey bee) was thought to have evolved in Africa. That idea now seems to be changing and the belief is that it originated somewhere in the Middle East. Whichever is the case, it spread to colonise all parts of Africa, all of Europe and the western parts of Asia, including Turkey, Iran, Saudi Arabia and other countries in that area.

During this movement, the original species changed to adapt to the alterations in climate and forage that it met in newly colonised areas. The result was the development of several different subspecies, each adapted to its own conditions and often separated from other types by some geographical feature such as a mountain range or expanse of water. However, they were still able to breed with one another, where they mixed, because they belong to the same species. Some of these subspecies are little known and not well documented but most have remained recognisably distinct.

Beekeeper interference

The subject is confused, as always, by beekeepers, who have moved bees about, taking subspecies from their natural locations and introducing them somewhere else so that there has been a great deal of mixing or hybridisation. (Incidentally, beekeepers have also introduced *Apis mellifera* into many parts of the world where it is not native, particularly the Americas, Australia and New

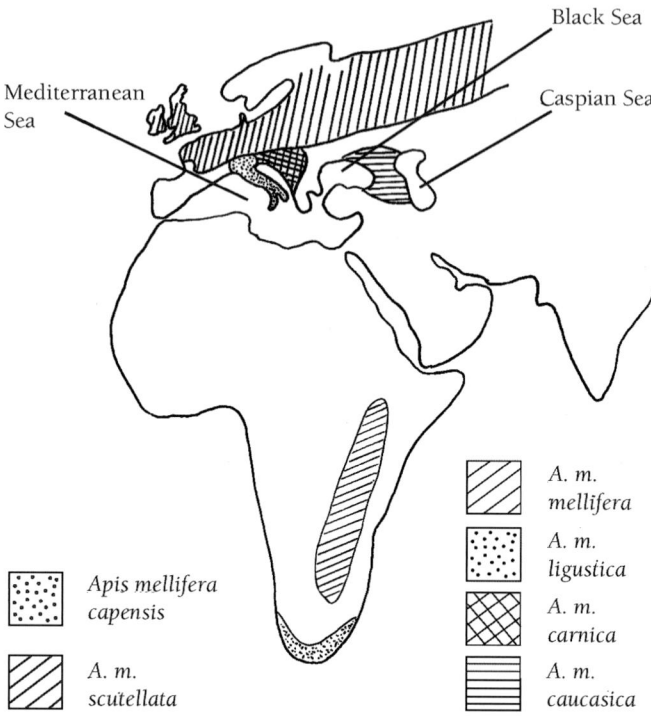

Location of some subspecies of *Apis mellifera*

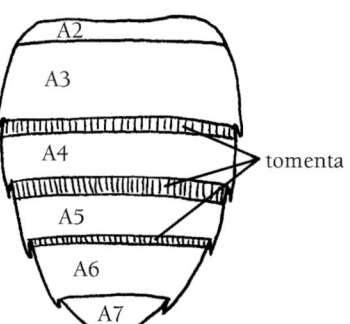

The width of the tomenta is one measurement used in determination of subspecies
A. m. mellifera has narrow tomenta, for example

Zealand.) Where subspecies have been moved into areas where other subspecies already exist, they have hybridised rapidly with the native bees so that, certainly in the UK, many bees are hybrids of very uncertain parentage and are often referred to as 'mongrels'.

Distinguishing characteristics

What criteria can be used to describe a subspecies? They are many and include appearances and behaviour as well as a number of specific measurements:

- size
- hairiness – bees inhabiting colder regions usually have more hairs to keep them warm
- the width of the tomenta, the bands of hair at the front of each abdominal segment
- colour – one of the most easily identifiable characteristics and, as well as general background colour we can include

here the colour of any spots and bands on the abdomen.
- tongue length
- measurement of various wing veins and the angles between them
- defensiveness/aggression
- honey production
- tendency to swarm.

It will be obvious to those of you who are still awake that some of these features, such as tongue length, are 'scientific' whereas others, like honey production, are 'beekeeping' characteristics and are therefore liable to beekeeper interpretation so, when we describe particular subspecies, we tend to use a mix of the two.

We must also remember that within a subspecies there may be considerable variation, particularly with respect to the 'beekeeping' characteristics, so that even though a subspecies is said to produce a lot of honey, one individual colony may be a much poorer honey producer than another. To complicate matters further, there is some natural crossing where the ranges of the various subspecies meet and, within a subspecies, there may also be recognisable groups.

The end result of all this is that we can only deal with broad classifications and tidy diagrams in books tend to be difficult to interpret on real bees, unless you are an expert.

GENERAL CLASSIFICATION OF SUBSPECIES

We can divide all the various subspecies of *A. mellifera* into three groups based on their natural geographic distribution:

- European
- Oriental
- African

This is self-explanatory and you will probably be relieved to know that I am only going to look at six subspecies in detail – four European and two African.

The main European subspecies

There are four European subspecies which I have placed in alphabetical order to avoid any suggestion of bias or favouritism:

CHAPTER 2

- *Apis mellifera carnica*
- *A. m. caucasica*
- *A. m. ligustica*
- *A. m. mellifera.*

Apis mellifera carnica (the carniolan bee)

The natural home of this bee is a huge area including the alps in Austria, east into Hungary as far as the River Danube and extending down into Slovenia, Croatia, Bosnia-Herzegovina and Serbia and Montenegro. It is greyish coloured and not too big. Its 'good' characteristics are:

- long tongue
- overwinters well on few stores
- resistance to brood diseases
- very gentle to handle.

So far, so good, but now we come to the disadvantages:

It builds up quickly in spring and then swarms. It loves to swarm! On the other hand, the queens are not very prolific layers and the bees do not like drawing out comb, so these would seem to be contradictions. Additionally, where there is a loss of nectar income for a time, breeding may cease altogether. Although the brood is resistant to brood diseases, the adult bees appear more prone to diseases such as paralysis, acarine and nosema. So, providing you can stop them swarming and keep the adult diseases at bay, this may be the ideal bee but beware because, as its natural home covers such a large area, there are many strains within the subspecies and these can vary tremendously in their characteristics.

Apis mellifera caucasica (the caucasian bee)

This bee evolved in the Caucasus, which is the mountainous region between the Black and Caspian Seas. It is in southern Russia going down into Georgia and Azerbaijan. The bee is similar in appearance to the carniolan bee with greyish colouring. Favourable attributes are:

- it is gentle
- does not have a very strong swarming instinct
- has a long tongue (probably the longest, on average, of all the European subspecies).

The problems arise with this bee when we consider overwintering

***Apis mellifera carnica* is native over a huge area**

and hive manipulations. You would expect a mountain bee to be good at overwintering and this one does not need much winter food, but it is susceptible to Nosema and so colonies tend to die from this disease. It is also slow to build up in the spring. Then there is its production of brace comb and use of propolis in great quantities. While these features may be of benefit to a colony living in hard conditions, keeping the colony snug in winter, they make hive manipulations much more difficult.

Apis mellifera ligustica (the Italian bee)

This bee originated in Italy, but there are various varieties; some are bright yellow in colour, but one, the Ligurian bee, is more the colour of leather. It comes from the Ligurian Alps in the north west of Italy near Genoa. Brother Adam believed that this was the best type of Italian bee and it was the one that was imported into Britain in large numbers after Isle of Wight disease destroyed so many colonies in the early part of the last century. In more recent years, the brighter yellow types have been imported into many countries in the world and many of the queens imported into the UK have been bred in New Zealand.

Apis mellifera ligustica
The Italian bee has been imported into countries all over the world

The Italian Bee is the most popular type of bee worldwide, is about the same size as *A. m. carnica* and it has got quite a lot going for it:

- it is very gentle
- can produce huge honey crops
- has a fairly long tongue
- is not prone to excessive swarming
- produces good comb.

But when it is taken into areas with longer winters and lower temperatures than those found in its homeland, it can have problems. These bees go through the winter with very big colonies, which need plenty of food. In the spring, in a country like Britain, they can find themselves in trouble because they have too many mouths to feed. Any food given to them by the attentive beekeeper is liable to be turned into brood very rapidly and these bees do not seem to have any idea of saving for a rainy day. Originally, part of the reason they were imported was their resistance to acarine but this seems not to be a strong characteristic today. They have a greater tendency to drift than other bees and they also are renowned robbers of other colonies. This may be seen as an advantage, providing the robbing colonies belong to you and the robbed colonies contain no disease.

CHAPTER 2

Apis mellifera mellifera
This bee is native to the UK

Apis mellifera mellifera

This bee evolved across northern Europe and into western Russia. It is found north of a line where the average July temperature is 15 °C or less and is the native bee of the British Isles. It has long body hair and is often called the British, or German, black (or brown) bee because it is dark coloured with very little lighter colour on the abdomen. Because its natural home is here, we might expect it to be the ideal bee for our conditions and, in many ways, it is. It is a large bee with a short tongue and has several characteristics which fit it very well for an existence in cooler climates with harsher winters:

- it does not breed at such a great rate as many of the other subspecies
- it overwinters well on few stores
- the individual bee is long-lived and very hard working with strong wings which enable her to fly further for crops
- they work longer hours than the yellow Italian bees
- they produce beautiful comb and good cappings.

On the minus side, they are quite strongly defensive (or aggressive depending on how you look at it), although proponents of this bee would question this statement. The low egg-laying rate means they do not build up strongly in the spring and honey crops are not huge.

Brother Adam stated quite categorically that this subspecies was extinct, killed off by Acarine, to which it was susceptible. However, it has always had its supporters and, recently, research using DNA techniques has confirmed that this bee is still around, probably in quite large numbers, although there has been a great deal of mixing with other imported subspecies. Oddly enough, it has been confirmed that this bee is present in Tasmania, Australia, where it was taken in the early part of the nineteenth century by settlers emigrating from North Yorkshire.

There are quite a few more subspecies in Europe and Asia, but we are going to look now at two African subspecies.

Apis mellifera scutellata **has developed characteristics which suit it well in its African home**

Apis mellifera scutellata (East African honey bee)

These insects are native to the south-eastern tropical parts of Africa and have a great many predators to contend with so they have become more defensive. They also benefit if they can remove themselves from unpleasant conditions such as severe drought so they have well-developed absconding behaviour. Add to this an overwhelming tendency to swarm frequently, usually supported by copious nectar flows, and you have the main characteristics of this subspecies.

Apis mellifera scutellata are small, yellow bees with short tongues and these were the bees which were imported into Brazil in 1956 and escaped to spread subsequently throughout parts of South America, all of Central America and the southern parts of the USA. They are still spreading. They are usually referred to as Africanised bees but are also sometimes described as 'killer bees', particularly in the popular press. They have become a major problem in areas that they have invaded, where they are the cause of a number of human and animal fatalities each year. They are an excellent example of the harm that can be done when subspecies of bees are introduced into totally new areas.

Apis mellifera scutellata **queen**
This subspecies has spread throughout South America and the southern USA as Africanised bees

Apis mellifera capensis (the Cape honey bee)

The natural home of this bee is the Cape region of South Africa, in quite a small area called the Fynbos, which has a very characteristic vegetation and quite harsh climate. *A. m. capensis* has evolved to succeed in this habitat. It was prevented from moving out of this area by mountains but, of course, beekeepers have achieved what nature, in its wisdom, never did and, in 1992 they moved these bees into other areas where they met *A. m. scutellata*.

The bees are dark in colour and have one other characteristic which sets them apart from *A. m. scutellata* and all other subspecies which have been studied. Workers can start to lay eggs in as few as six days after the loss of their queen and these eggs can develop into females. The eggs are not fertilised and this method of reproduction is known as *thelytoky* (production of females by parthenogenesis). It is thought that, in this instance, the mechanism underlying thelytoky is the recombination of two of the nuclei produced by the meiotic division. So meiosis occurs normally in the development of the egg, with the production of four nuclei, but is followed by the fusion of two of the nuclei to give a diploid egg, which then develops into a female.

This might seem to be an advantage to the individual *capensis* colony in the case of queen loss but, where they have been mixed with *scutellata* colonies, *capensis* workers moving into the *scutellata* colonies are not controlled by the pheromones of the incumbent queen and rapidly produce eggs which develop into pseudoqueens. The production of many pseudoqueens from these eggs causes anarchy and chaos, the *scutellata* workers rear the *capensis* bees, which are very poor foragers, and eventually the colony dies. *A. m. capensis* effectively functions as a parasite or cuckoo in the colonies of *A. m. scutellata* and the so-called 'capensis problem' has resulted in huge losses to beekeepers in southern Africa. The problem is ongoing.

The future

From the biological point of view, it would seem to be best if subspecies of *Apis mellifera* remained in their own regions but, beekeepers being what they are, bee types have been moved about and many managed bees today are hybrids. One of the common effects of hybridisation is aggressiveness, even where the parents come from docile stocks. The precise reason for this is not known, but the various subspecies do have different 'dialects' with respect to pheromones and other forms of communication and this may be part of the answer. Another view is that defensiveness is a good biological trait. ie, bees showing defensive behaviour will be better able to compete in the world, and that it is a form of hybrid vigour achieved when the genes from two distinct subspecies mingle. In all likelihood it is probably a combination of factors.

Some attempts have been made to breed hybrid bees which combine the good characteristics of several different subspecies, as is done with domesticated animals, and these have sometimes met with limited success. An example is the 'Buckfast' bee, which

Brother Adam developed as a result of his observations on bee subspecies over many years and in many parts of the world. The snag is that such a bee is a result of mating particular, carefully selected lines and it is impossible for the ordinary beekeeper to repeat it. So, once this route has been chosen, it becomes necessary to keep purchasing the queens from elsewhere to maintain the strain, otherwise the young queens will mate with local drones and problems will often ensue.

As is clear from the examples of Africanised bees and of *A. m. capensis* detailed above, the introduction of 'foreign' bees into a region can cause enormous damage to the bees and beekeeping in that area and should, therefore, always be done with caution and with regard to the wider implications, while remembering that the greener grass on the other side of the fence may become less palatable as it ages. Having said all that, there is the urge to improve the stock we have, so we will now take a look at breeding systems and stock improvement.

ONLY THE BEST

With the number of honey bee subspecies, and the problems that can occur when bees are moved about and interbred, bee breeding seems fraught with difficulty, but that does not mean that nothing can be done to improve our bees. You need to ask yourself three questions: Do you have the bees you want? Do they do everything you want them to? Do they do a few things which you really wish they didn't? Few honey bee colonies are perfect, but improvements can be made, as in any form of livestock, by selective breeding. So, how to go about it? There are two ways:

- rely on someone else to do the work for you and simply buy mated queens from them
- do the job yourself by rearing queens (intentionally!) from particular colonies with good characteristics and (equally important) culling those queens that do not meet your expectations.

Let's deal with the first idea.

If you buy mated queens and introduce them to your colonies you will notice improvements. Bought from a reputable dealer, a lot of hard work will have gone into raising those queens. But what are the future prospects? Those queens will produce, either by design or in their own natural way, daughter queens. These will mate with local drones and then bang goes your carefully selected stock. Often these hybrids are very bad tempered or have other undesirable characteristics. Some of the worst-tempered bees I have seen were the colonies produced by the second generation queens of docile, gentle, Italian-type, purchased queens. So where does that leave you? The obvious answer is to buy a queen whenever you need one and, for the beekeeper with perhaps two colonies, this may be the best way forward. It also helps to keep the bee breeders in business but, because these queens produce drones which mate with the local queens, your beekeeping friend down the road may not be so happy with his hybrid queens.

So to Plan B – breed your own.

Desired characteristics

Before you do anything you need to make up your mind what characteristics you are looking for. Below is a list, which is not necessarily complete but covers the main points.

- *docility*. This must surely be top of everyone's list because gentle bees are so much easier to handle than very defensive ones and are less likely to cause problems with neighbours or the general public. In the case of the urban, and suburban, beekeeper, and in these litigious times, docility is an absolute must.
- *disease resistance*. Vitally important because nothing is so dispiriting as colonies which are always sick and struggling or, worse still, those that have to be destroyed. Also, colonies with disease problems never realise their full honey-producing potential. Remember that some bees are more likely to suffer from a particular disease because of the genes they inherit from their parents.
- *ability to produce lots of honey*. You would think that this would be of major importance to everyone but some beekeepers just want enough honey for family and friends and do not want the mess and trouble of extracting masses of the stuff.
- *tendency to swarm*. Undoubtedly some strains of bee swarm more than others, but I believe that the colony which builds up well in the spring, gathers lots of honey and then prepares to swarm in May or June, is acting in a sound biological way. It is an unfortunate fact of life that bees and beekeepers do not work to the same agenda, so success for the bees does not necessarily mean success from the beekeeper's point of view! It is possible to select from those colonies with a low

swarming tendency but I believe that swarming can be seen as a husbandry matter unless a colony starts to swarm when it is quite small or tries to swarm twice in a season. That is unreasonable and becomes a matter for selection.

The list of other characteristics, which we could consider, is long. Some of them are reflected in the four points above, such as fecundity (number of eggs the queen lays), average length of life of the workers, and hygienic behaviour (ability of the workers to detect and remove diseased brood). Others, such as colour, are not. Then there are those individual concerns, for example, if you produce a great deal of comb honey for sale, bees which cap nicely with uniform, white cappings, are a real asset, bees which do not produce too much propolis can be much easier to work, and so on.

So, having drawn up our list of desirable characteristics, what do we do next? In most cases, absolutely nothing! But we can improve things, sometimes in surprisingly simple ways. How do we go about it?

Know what you have

To know what your colonies are doing, how much honey each produces, when they swarm, what disease they have and what their temper is like, you *must* keep records. Otherwise you are relying on your memory – enough said!

Even with one or two colonies, simple records are essential for the general upkeep of the colonies and, by selecting the correct criteria, these same records can be used for selection of breeding stock. That sounds grand doesn't it? The type of record sheet/card is immaterial, as long as it can be taken with you to the bees, filled in simply and quickly on the spot and is easy to understand when you get home. I use a plastic loose-leaf file with A4 sheets run off on the computer, one for each colony. Over the years, the column headings have changed and my daughter and I have our own shorthand for filling it all in. It takes perhaps half a minute to find the page and fill in each sheet as soon as we have closed the hive. By the end of the season, the sheets are smeared with propolis and are sometimes stuck together with nectar, but the system works.

Marking queens is the single most useful step a beekeeper can take towards bee improvement and better husbandry

Mark your queens

I think that marking his/her queens is the single most useful practical step any beekeeper can take to improving his/her beekeeping. If you intend to do any selective breeding you must be able to recognise, and find, each queen. It is also essential for many husbandry procedures but that is another matter. A simple dab of paint will do, although you can use numbered discs which are glued to the thorax.

Your records will show you how old the queen is, who her mother was and how she was raised. Separate to your hive record, you can build up a family tree for queens.

Be objective

When selecting a queen to breed from, do not be swayed by emotion. We all have the queen who looks good, the one we see striding majestically over the frames every time we open her hive. We get quite fond of her. But, if she does not match up to your criteria, do not use her as a breeder queen. Select the best you have for the characteristics you want and use her or them.

See what the kids are like

In science-speak this is called progeny testing. What it means is that, before selecting a breeder queen, you should see how her daughters perform. Of course it is no good looking at the daughters' performance if mum has long since disappeared over the hills with a swarm, so swarm control is essential. A promising queen which is getting a bit long in the tooth can be kept going in a nucleus where her energies will be conserved and she will be safe.

RAISING NEW QUEENS

Having decided which of your queens most closely matches the characteristics you want (and none will be perfect), you need to raise new queens from her or them.

I am not going into queen rearing methods here as there are so many options, but the very simplest is to take queen cells from your desired stock when it makes swarming preparations. You can do this even if you cannot find queens or be bothered with queen rearing the proper way! Put them into nucleus colonies and, when they are mated, introduce them into the colonies you want to requeen. Of course, at this stage, you will have to find queens.

If you want to rear queens in a more controlled fashion, it is not difficult as long as you remember a few basic rules:

Queen cells produced by grafting day-old larvae into artificial wax cups

- a stock of bees left queenless for about four hours will begin to raise new queens from young female larvae
- if you give newly hatched larvae (less than 24 hours old) to such a colony, ensuring that it has no queen cells of its own, they will rear queens from these
- to rear good queens, a colony needs a large number of well-fed, young nurse bees and plenty of pollen which must be available for several days before the young larvae are given to the colony
- mature queen cells must be transferred to mating nuclei one or two days before they hatch
- queen cells must be handled with care at all stages – and never shaken.

Sealed queen cells produced by a Jenter system

How you get the newly hatched larvae into the colony is up to the individual. They may be grafted, produced using a Jenter kit or similar, punched from cells, or there is a variety of other methods. What is vital is that they must be no more than a few hours old.

Curtains for some

An essential part of stock improvement is to weed out the weak ones. So, while concentrating on breeding from the best, you must remove the worst. I know beekeepers who cannot bear to kill a queen, but you must put all feelings aside and wield the axe (not literally – a finger and thumb is quite sufficient) if you want to make progress.

What about the dads?

They are a problem. We can control our queen raising but it is difficult to control mating unless you resort to II (instrumental insemination). Remember that the drones your young queens mate with are just as important as the queens themselves and probably the best you can do is to ensure that you have no colonies with characteristics which you do not want, then persuade all the beekeepers in your neighbourhood to apply the same principles. Of course, you will all have to agree on what you want from your bees and control swarming and cull queens that do not come up to scratch and you will probably decide that breeding flying pigs is an easier bet but, even on your own, you can improve things and, who knows, when your fellow beekeepers visit your apiary and see what good bees you have, they may be inspired to follow your example.

CHAPTER 2

MORE ADVANCED BEE BREEDING

Honey bees are not like cows! 'Well of course they're not,' I hear you say. 'Cows have only got four legs'. My other reader can probably think of other differences, but there is one big similarity: both cows and honey bees are used by man to provide useful products which, incidentally, are often spoken of in the same breath – milk and honey. Once man starts to keep an animal for production there comes the wish to improve it so that its production increases, and beekeepers are no exception. We have seen how we individual beekeepers can select good stocks, breed from them and eliminate the poorer queens. Now I want to look in a bit more detail at breeding systems and this is where our cows come back. To improve a particular characteristic, such as milk production, the breeder of dairy cows selects, from the animals available to him, those cows which have the highest yields. (He will have to look at other characteristics too, but I am trying to keep things simple.) He will choose bulls, which also come from high-producing mothers, and mate these to his good cows. He will then keep selecting through generations of animals so that the milk yield gradually improves. It is very easy to control mating in cows and there are different ways of doing this:

Line breeding

Line breeding involves using a breeding group to create a 'line'. In each generation, parents are selected from the same group by choosing those showing the desired characteristics, in this case high milk yield, and so on, in succeeding generations. Line breeding can result in *inbreeding* where very closely related animals, such as brother and sister, are mated, but it will depend upon the size of the population at the beginning. A larger population will give more choice of parents.

Cross breeding

Cross breeding is the precise opposite of line breeding and uses totally unrelated animals at each cross. They may be animals from different lines, which have been developed by line breeding. Using this method it is often possible to combine various good characteristics from different lines so that the final result is a far superior animal. We call the animals produced by cross-breeding

An isolated mating site

hybrids and in genetic-speak they make up the *F1*, or first filial, generation. They are usually a uniform group of animals with very similar characteristics. Although the hybrid may be a very good animal, it is necessary to go back to its original parents, or at least to the original lines, to repeat it. If the hybrids are allowed to interbreed, giving rise to the *F2* (second filial) generation, a very variable group of animals will result, some good, some poor.

Bees are more of a problem

Where does this get us with bees? To recap on mating behaviour of the honey bee: a newly-emerged queen mates with up to about 20 different drones, on average, during the first three or four weeks of her life. Each drone dies immediately after mating and the whole process takes place between 10 and 40 metres from the ground and some distance from the hive. It is apparently impossible to induce natural mating under controlled conditions so the bee breeder's life is much more difficult than the life of a dairy cow breeder. The whole idea of the mating system in honey bees is to enable random matings to take place, to ensure that a drone is only used once, that matings of closely related individuals are kept to a minimum and the maximum variety of genes is introduced into the colony.

So, how do we try to control mating and are there any problems related to doing this?

A mating nucleus situated at an isolated mating apiary at Spurn Point, Yorkshire

Isolated mating

Isolated mating stations have been tried and are in use in various places. The idea is to find an area, such as an island, which contains no other honey bees and is sufficiently far from other colonies to avoid contamination. Selected colonies containing many drones from colonies with the desired characteristics are placed in the area, the virgin queens are taken there and nature is left to take its course. This is not foolproof of course as other drones can, and do, find their way there, but it gives an acceptable level of control.

Instrumental insemination

Instrumental insemination (II) seems to provide a life-line, enabling bee breeders to control the mating of their bees. Amazingly, bee II has been around for more than half a century so it is not a new technique. With the ability to control mating, all the principles

An anaesthetized queen ready for insemination with her sting chamber opened

that the cattle breeder used can be applied to bees but there are problems, principally the skill involved in learning the technique, the time involved and the cost of the equipment. All of these put it out of reach of the average beekeeper.

Because a honey bee is small, a microscope is used to see what is happening. The queen, held firmly in position, is anaesthetized with carbon dioxide and receives a quantity of semen which has been collected from the desired drones. Very fine instruments are needed and scrupulous cleanliness must be observed. The queen is returned to her nucleus hive and care must be taken that she does not go out on natural mating flights. Then there is the efficiency of the matings. Do the queens lay as well or for as long as naturally mated queens when introduced to a production colony? I have no personal experience of II queens so I do not know the answer to this question but there are constant mutterings about the queens' reduction in laying and useful life, so it may be that queens produced by this method need replacing every year. However, with modern techniques, it should be possible to get as much sperm into a queen's spermatheca as with natural mating. Semen from a number of drones can be mixed, to simulate the natural situation and then many of the problems can be alleviated. The real value of II, at the present time, is as a tool in the hands of professional queen breeders and research workers.

Even if the problem of random matings can be overcome, there are other snags to applying the breeding systems which have been used with such success in cows.

Line breeding in honey bees goes against all the aims of natural bee mating systems. It leads to inbreeding which a million years of evolution has abolished. Because of the special nature of sex determination in honey bees, close matings can result in the production of diploid drones, with a corresponding loss in worker bee population which may be as high as 50%. The degree of inbreeding will depend on the size of the initial population, ie, the number of colonies at the beginning of the process. Maintaining lines of bees is difficult and requires a great deal of skill.

Cross breeding to provide hybrid queens, using established lines, seems to give the best prospects. The resultant crosses (F1) should combine the good characteristics of both parents and show hybrid vigour (just like a mongrel dog). Many of the well-known queens produced for sale commercially, including Buckfast, have been hybrids. The disadvantage is that to obtain uniform queens, the cross has to be made every time, so lines have to be maintained and II usually has to be used to achieve this. Clearly a job for the professional. Where beekeepers buy hybrid queens and then allow them to produce daughters (F2) which mate naturally with local drones, the results are very variable and can lead to all sorts of problems, the most obvious one of which may be extreme defensiveness.

CONCLUSIONS

To go down the route of controlled breeding of honey bees needs dedicated breeders who will produce top-quality queens with good characteristics to sell to the ordinary beekeeper. In some European countries such schemes exist and beekeepers are happy to buy their queens on a regular basis. Using such a scheme, the National bee colony could be improved in the same way as the National dairy herd was improved, but this is Great Britain!

The individual can work in this way on his own and increase his honey yields (as well as maintaining other desirable characteristics) but it seems unlikely that the principle could be extended on a wider field. We are a nation of hobbyist beekeepers, very few make their living from bees and, in my experience, any group of 10 will always have at least 11 solutions to any one problem, so how are we ever going to agree on the ideal queen? Indeed, it is unlikely that one type of queen would serve all beekeepers from Cornwall to Scotland so there would have to be many regional variations. There is also the question of cost. How many beekeepers do you know who will buy new queens, on a regular basis, when they, or rather their bees, can rear perfectly 'good' ones themselves?

There is a glimmer of hope – the reduction in the number of bee-owners and feral colonies, due to the effects of the varroa mite, means a more committed beekeeping community and, of course, if some breeder out there produces a naturally varroa-resistant bee while maintaining other good characteristics, then who knows?

3 HEALTH AND HAZARDS

KEEPING THEM HEALTHY

It is easy to start any discussion on bee health with a list of diseases and disorders. This is not only confusing but also very discouraging, but there is a more positive aspect to the maintenance of healthy colonies, so that is where we will begin. Then, in the second part of the chapter, we will look at the various types of disease-causing organisms which we might find in our bees.

Many diseases arise because disease-causing micro-organisms (pathogens) have invaded their host and multiplied within it until there is noticeable illness, but to assume that the presence of a pathogen equals disease is an oversimplification of the matter.

In our human world we strive to eat a balanced diet, to keep our homes, and ourselves, clean and we now know that stress can be a contributory factor in many diseases. We recognise that spending time with a lot of other people crowded into warm places favours the spread of any infections and we know that disease prevention involves keeping healthy so that we can withstand infection, preventing pathogens from entering our bodies in the first place and, only as a last resort, destroying the pathogens that are causing us to be ill.

A crowded beehive is an ideal place for disease organisms to flourish

So why should bees be any different?

Think about the conditions inside a beehive. It is crowded with thousands of little bodies, it is always warm, there is plenty of moisture about and pathogens are ever present. This is the first fact to appreciate – the micro-organisms causing most of the common bee diseases are probably present in every hive (just as many human pathogens are present in a room full of people). It is the activities of the bees themselves which keep the pathogens under control, but

Bees benefit from minimal interference but it can be taken to extremes!

the beekeeper can do a lot to help – or, more likely, hinder.

Bees have good immune systems, both individual and at a colony level, which enable them to combat pathogens and maintain colony health and these need supporting.

Not much is written about the immune system of the individual bee but it operates at several levels including a chemical level within the haemolymph. Bacteria in the gut of the bee and in bee bread produce some natural antibiotics, which help to combat pathogens invading the gut. Use of antibiotics in the hive can kill these bacteria. At a colony level, propolis is a vital component of the bees' protective armoury and we should remember this as we remove it (see Chapter 10).

The natural order

If we go back to basics and think about the natural cycle of colony life in the bee world, then compare it with bee life in an apiary, we can begin to see why we have such problems with disease.

A colony will start life as a swarm, in a new cavity, building new comb. All nice and clean, although the bees will bring some pathogens with them. The colony will be some distance from its nearest neighbours and, if it is lucky, it will gather sufficient food to survive the winter so that the next year it can swarm again, perhaps more than once. Always a proportion of these swarmed colonies, and the swarms, will fail to survive, for a variety of reasons. So the bees die out, stores, brood and the comb are all destroyed rapidly by wasps, mice, wax moths and many other animals which appreciate such a bounty, and in the process, pathogens are destroyed. Even if the cavity is colonised by another swarm, it will be clean and the new colony will cover the inside of the cavity with a protective coating of propolis.

But compare this with the situation in a managed apiary.

There, the beekeeper will keep a number of hives cheek by jowl, give old comb to swarms, nurture ailing colonies or combine them with other colonies and, if a colony dies out, (s)he will keep the comb for future use. (S)he will also feed the bees with artificial food, shut colonies up while they are moved about in a bumpy trailer and tear the whole lot apart at regular intervals.

I could go on but you get the drift – beekeeping, from the bees' point of view, is an artificial situation. That does not mean that we should not manage our bees. Leaving them alone is not an option and we must recognise that, once we have put them in a box, they are our responsibility.

So, how can we keep our colonies healthy?

HEALTH AND HAZARDS

Sound beekeeping

Many of the practices which help to keep disease at bay are simply good husbandry. They prevent the movement of pathogens or increase the colonies' chances of fighting off any infection:

- ensure that colonies are headed by young, vigorous queens. They can often out-breed any infection
- replace queens in colonies which seem particularly prone to develop disease, with queens of different strains
- take great care when moving comb from one hive to another
- never feed honey, or pollen, from an unknown source
- never unite weak colonies with others until the cause of the weakness is known
- isolate stray swarms and monitor them for disease
- hive swarms on new foundation
- if you buy bees on combs, inspect them very carefully, or get assistance.

Keeping it clean

Cleanliness is said to be next to godliness and is certainly of vital importance:

- if you buy second-hand equipment do not allow bees access to it until it has been cleaned thoroughly. Treat it as if it was covered with long-lasting, disease-causing spores – it may well be – and sterilise all boxes with a blow-torch
- never use second-hand comb
- have a regular programme for frequent replacement of used comb and frames. Disease organisms survive in comb and on frames.
- remove squashed bees so that other bees, in cleaning them up, do not become infected from their remains
- be tidy around the apiary. Avoid leaving old comb and bits of wax lying about. This can start robbing
- fumigate all out-of-use brood combs and boxes with 80% ethanoic (acetic) acid. Have a system for removing each brood box from use for a time to enable this to be carried out. The acid will destroy almost all of the disease-causing organisms which lurk on surfaces inside the hive.

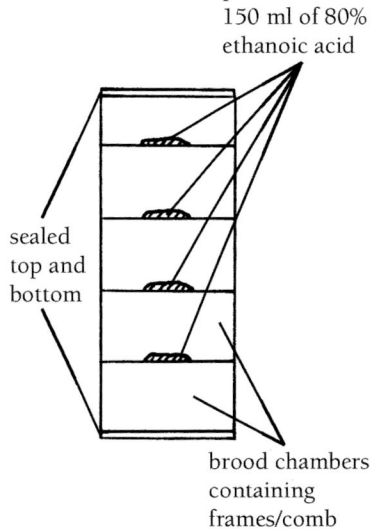

Fumigation with 80% ethanoic acid

Avoiding stress

We all know what we mean by stress in humans but stress from

Migratory beekeeping, such as taking bees to the heather, can cause stress

the point of view of the bees can be anything which renders the colony more susceptible to disease arising from the pathogens already present in the hive. The list below is long, and it must be appreciated that some stressful situations may be unavoidable, but even if this is so it helps to be aware of them.

Try to avoid:

- anything confining bees within the colony
- too frequent handling and manipulations, particularly at inappropriate times (cold, wet, windy)
- poor supply of pollen, particularly in the spring
- high density of colonies in an area
- migratory beekeeping
- high concentration of carbon dioxide in a hive
- presence of any disease or infestation
- damage from pesticides
- keeping bees in unusual situations such as poly-tunnels.

Some of the items on this list are connected: migratory beekeeping, in addition to confining the bees within the hive for periods of time, often also increases the concentration of colonies in an area, with consequent reduction in foraging activity. Poor weather and crop failure belong in the category of factors confining bees in the hive, both generally being out of the beekeeper's control. Keeping bees in situations such as poly-tunnels will restrict their foraging range, reduce the range of pollens and nectars that they can collect and generally slow colony development. A good supply of mixed pollen is essential for healthy larval growth. Bees may have an abundance of pollen but, if it is all of one type, it may well lack some essential amino acids (the building blocks from which proteins are made).

Factors which unbalance the hive population, resulting in fewer house bees, can result in poorer cell cleaning and survival of more pathogens. Spray damage falls into this category and this can also lead to reduced foraging ability and shortage of food.

Small colonies of bees can find it more difficult to ventilate the hive and levels of carbon dioxide may rise. This can happen in queen-mating nuclei which also suffer from an imbalance of workers and constant disturbance.

The presence of one disease can lead to the development of others as the weakened bees succumb to other pathogens. We also know that some pathogens, particularly varroa and *Nosema ceranae* depress the bees' immune systems.

Looking at this list it is obvious that normal beekeeping practice will, unavoidably, make colonies stressed. The aim should be to keep it to the minimum.

Dead bees outside a hive after being affected by poison

HEALTH AND HAZARDS

Monitoring for disease

It is important to know what, if any, diseases are having noticeable effects on the colony and there is no easy way round this other than to look for them. You need to:

- look regularly for brood diseases. At least once a year, shake the bees from each frame so that you can see the brood clearly. Remember that disease may be visible only in a very small number of larvae. Remove cappings from cells which look doubtful. (A cocktail stick is a very useful tool for this.)
- monitor constantly for varroa (see Chapter 6). Any increase in varroa mite levels will lead to increased susceptibility to many other pathogens
- where colonies are slow to build up in spring, examine microscopically for adult bee diseases – or get someone else to do it for you (see Chapter 5)
- every time you inspect a colony think 'disease' and look for anything unusual, particularly a poor brood pattern with many gaps, abnormal larvae and dodgy cappings.

Know what is normal

The most important knowledge a fledgling, or any other beekeeper, can have is to recognise healthy brood. (Adult bees are a different matter and are usually either alive or dead. They rarely look poorly.)

Unsealed brood. A very young larva is white, shaped like a letter 'C' and lies at the bottom of its open cell. It grows rapidly, is soon shaped like a letter 'O', but is still a glistening, pearly white. Its body is divided into a number of sections called segments and these are clearly visible. When it is fully grown, the larva fills the cell but is still curled round like a very fat, and solid, letter 'O'.

Healthy unsealed brood

CHAPTER 3

Healthy sealed brood

Sealed brood. Once the larva is fully grown, the workers close its cell with a porous wax capping and it is then called sealed brood. It will remain like this for 12 days if it is a worker. Within the cell, the larva stretches out so that its head is pointing towards the cell capping. It then changes into a prepupa and, after this, into a pupa. The capping above a healthy prepupa or pupa is slightly domed, of a light brown colour and with a rough, dry texture. There should be no visible holes in the capping, but remember that it takes time for the worker bees to construct a cell cap and a fully-developed bee does not chew its way through its capping instantly, so a hole may be seen at these two points.

The brood, as a whole, should be in fairly solid slabs with only a few empty cells. A large number of empty cells indicates that larvae/pupae have died and been removed by the bees.

Once the beekeeper can recognise healthy unsealed and sealed brood, it is easy to see a potential problem. Then it is a case of investigation, consulting the books or, if (s)he is a novice, screaming for help from those more experienced.

TYPES OF ORGANISM CAUSING DISEASE IN HONEY BEES

Various different organisms, many of them very small, can cause disease in honey bees and it is useful to understand a little about them.

Viruses – the hidden enemy

Our world is full of viruses. They invade living cells of all plants and animals, including us, and some, such as HIV and some types of influenza, can cause mass panic, major epidemics and major concern to governments. Fortunately, most stick to one host although there are exceptions, such as the rabies virus. They are often present in an organism without causing any noticeable disease.

All viruses consist of a small amount of genetic material (RNA or DNA, but only rarely both) contained in a coat made of protein but, although each contains the blueprint for making more viruses, they are extraordinary because they can only reproduce when they are able to 'hijack' another living cell, making use of the host cell's ability to produce energy and build proteins. As a result, the host cell is programmed to produce new, fully formed and infective viruses, called *virions*, and the normal functions of the cell are shut down.

If the virions remain inside the cell usually no problem arises, but if they escape from the cell, or kill it, they move on to infect other cells. Once sufficient cells are affected to cause visible signs, we recognise a disease condition. The individual virions are very tiny and can only be seen with an electron microscope, but in some viruses, particularly those infecting insects, many virions aggregate together and become embedded in a crystalline structure. These structures are characteristic of each virus and can usually be seen with a light microscope.

Because a virus can only actively live and reproduce inside the living cells of its host, and will usually die very quickly outside its host, it is called an **obligate parasite** and it is very important for it to be able to move quickly and easily from one individual to another. Viruses are helped in this if large numbers of their host live in a confined space, and are aided by food sharing between individuals. Their spread is rendered even more efficient by the presence of other parasites which suck blood and act like tiny hypodermic syringes, injecting viruses into their hosts. (When parasites act in this way they are called *vectors*). The effect of the myxomatosis virus, which is spread by fleas in the rabbit population, illustrates this point perfectly as well as demonstrating the enormous damage that can occur where a host species has no resistance to a particular virus.

Another characteristic of viruses is their ability to mutate, that is change the structure of the RNA/DNA. Such changes give rise to a number of different forms of the virus, some of which may be more virulent than others. We are all familiar with this idea from the influenza virus which is continually mutating to give different forms which can rapidly cause epidemics.

We have a further method of spread of viruses which is called *vertical transmission*. This means that, rather than spreading from one individual to another, some viruses may also be transmitted from queen to egg, sperm to queen and so on. So, an individual bee may be infected from the egg onwards.

To a honey bee virus, a bee hive must be paradise and when an organism such as **Varroa destructor** appears on the scene, to act as an efficient vector, it must seem like its Birthday and Christmas rolled into one.

Bacteria – small but often not beautiful

There are huge numbers of bacteria in the world both in terms of species and actual numbers. The problem is that they are so small (but not as small as viruses) that a microscope with a magnification of a least x1000 (one thousand times) is needed to see them, so most of us are pretty unfamiliar with them. Bacteria are basically unicellular (each one is made up of only one cell) and they are described as **prokaryotes** (a good word for impressing pompous acquaintances!). It means that a bacterial cell does not have a discrete nucleus to contain the DNA (genetic material) and lacks some structures which would normally be found in a nucleus and in the cytoplasm of higher organisms (eukaryotes). Bacteria can be various shapes such as rods, spheres or spirals, and may move about, using a flagellum (a long, beating hair-like structure), or remain static. In some species the individual cells can form aggregations so that they may appear as clumps or chains. Bacteria reproduce, at an enormous rate, by simply dividing in two and some can produce very resistant spores which can survive outside the host in unfavourable conditions, but still remain infective, for a very long time.

Bacteria are found everywhere and many of them are essential to the correct functioning of plants, animals and the world in general. Some are found in the guts of animals, including us and bees, to help with digestion; others use atmospheric nitrogen and put it into the soil in a form that higher plants can use; many are concerned with decay so that organic matter can be re-used by other organisms and, in this respect, some bacteria are used in the breakdown of sewage. A few produce antibiotics. On the minus side are the baddies which cause disease, mostly in animals but also in plants. Among these are the bacteria causing anthrax (the first to be discovered), tuberculosis and pneumonia. Many infect insects.

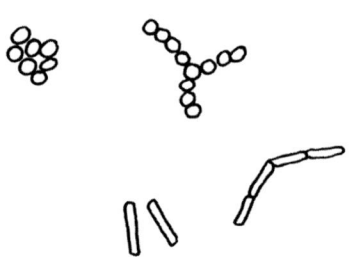

Bacteria
The single cells may be round (cocci), rod-shaped (bacilli) or other shapes not shown. They often form clumps or chains (magnification x1000)

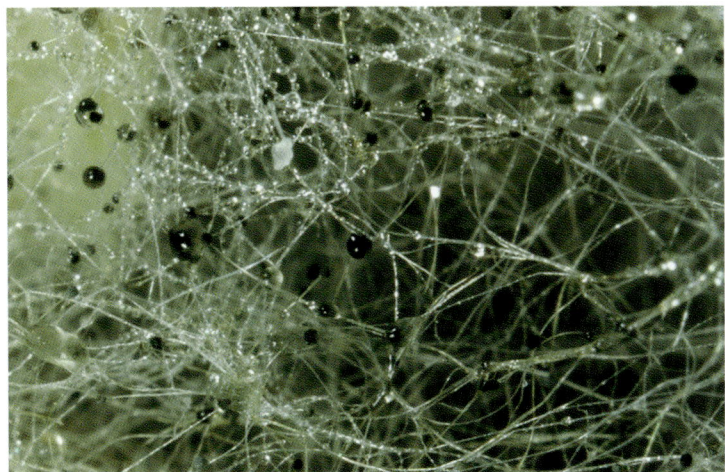

A saprophytic fungus growing on bread, showing a network of hyphae. The black structures contain spores

Fungi

Fungi are a huge group, forming a separate Kingdom. Unlike the bacteria, their genetic material is contained in a nucleus. They are therefore a group of *eukaryotes* (organisms having the cell nucleus surrounded by a nuclear membrane) but they often do not have any cell walls so a number of nuclei just float in cytoplasm. Most form long threads called *hyphae* (singular: hypha), which can develop into large networks or solid structures like mushrooms. We are not so familiar with fungi as disease-causing organisms in the human world, unless we suffer from Athlete's Foot, but many are parasitic on plants, being responsible for such horrors as potato blight and the various rusts, and some attack insects and other animals. Many of them are saprophytes, growing on dead organic matter. Here they perform a very useful function, which can sometimes get out of hand as far as we are concerned (think about mouldy food and dry rot), as well as providing us with some culinary delicacies such as truffles.

Microsporidia

All microsporidia reproduce inside animal cells and are therefore obligate parasites. There are probably over one million species, all with their own hosts, and they infect every major group of animals. They are simple organisms, classified as a separate Phylum within the Kingdom of Fungi and, although they are eukaryotes, with well-defined nuclei, they lack some other cellular structures. They have a characteristic, and unique, way of infecting cells by 'firing' a coiled, hollow tube into a cell. The microsporidian cell contents then pass through this tube into the host cell. The whole apparatus is a bit like a harpoon with a hypodermic needle on it. Microsporidia frequently infect the gut in their hosts, often insects, but may attack other organs and some species can even cause severe problems in many organs in humans with compromised immune systems, such as those suffering from HIV.

Protozoa

Protozoa are single-celled animals found in all kinds of damp or wet places. The single cells may be aggregated into colonies. Among the Protozoa is the well-known *Amoeba*, which slides around in muddy ponds and about which we all learned at school, many saprophytes and free-living animals, as well as some causing disease such as the organism responsible for malaria. Reproduction is frequently achieved by simply splitting in two, so it can be rapid.

Mites

These are bigger than the organisms we have considered so far. Their place in the grand order of things can be shown in a simple diagram:

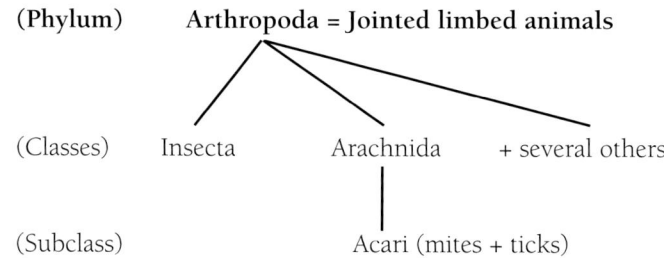

This is a simplified, and controversial, classification as a number of sub-phyla between Phylum and Class are often inserted and some classifications do not classify Acari as a sub-class but as an Order or super-order. The Arachnida (anglicised version Arachnids) includes all those creatures which have two sections to their bodies:

- a front section called the prosoma or cephalothorax (head + thorax) carrying two pairs of appendages, making up the jaws, and four pairs of legs, although in mite larvae there are only three pairs
- an opisthosoma or abdomen, which carries no limbs but may have an appendage at the end of it.

HEALTH AND HAZARDS

The Arachnids which are most familiar to us are the spiders, but there are at least another ten Orders within the Class. The scheme of classification shown demonstrates the close relationship between the Arachnida and the Insecta and this suggests that controlling mites in a colony of bees might be slightly difficult. Mites are a very numerous group with around 50,000 known species in the Acari and many more as yet unknown but, as they are all small, most microscopic, they are not very noticeable. Many are parasites, both inside (endoparasites) and outside (ectoparasites) the bodies of other animals and plants but, as beekeepers, we are only interested in a very small number.

An incomplete list

That is a fairly formidable list of different types of organism that can attack our honey bees and it is not even complete because I have not included some of the larger pests and predators here (see Chapter 7). In the next chapter, we will start to look at the diseases themselves.

4 WHEN THE KIDS ARE ILL

Like most youngsters, honey bee larvae suffer from a variety of ailments. Most of these kill the individual larva but, on a colony scale, some of them are comparatively minor while others are very serious and can destroy the whole colony. The signs of each condition seem clear cut when we read about them or look at photographs, but many are far more confusing when we are faced with them in a full colony. I am afraid that there is no alternative but to go through them one by one, following the same scheme as that adopted in the previous chapter.

VIRUSES AFFECTING BROOD

Beekeepers largely ignore viruses. There are good reasons for this:

- many beekeepers know little about viruses
- the names of the viruses are not very memorable
- diagnosis is difficult and frequently combined with signs from other non-viral diseases
- treatment is impossible.

With the constant presence of varroa, more beekeepers have become aware of viruses and the problems they can cause. There are many viruses affecting adult bees and we will look at these in the next chapter, but there are two brood diseases caused by viruses, one quite common and relatively easy to recognise, and another less well-known and affecting only queens. We will start with the easy one.

Sacbrood

The sacbrood virus gets into a larva when it is young, usually about two days old, but it does not kill the larva until after it is *sealed* and

CHAPTER 4

Sacbrood
The Chinese slipper effect can be seen in the cell in the centre

is at the prepupal stage. The virus causes the fifth moult (when the prepupa changes into a pupa) to go wrong. The tough outer cuticle is not shed and the space inside it fills up with ecdysial, or moulting, fluid. This is the fluid produced during the moulting process and it separates the outer cuticle from the living layer of cells, the epidermis, underneath. So the prepupa, instead of changing into a pupa and continuing its development, becomes enclosed in a *fluid-filled 'bag'*, or *sac*, and hence the name of the disease. Not surprisingly, the prepupa dies. It *changes from a yellowish colour to a dark brown*, which spreads from the head downwards, and it quickly dries into a dark brown flattened scale which is shaped like a gondola (or a Chinese slipper, depending on your knowledge of these two objects), ie, gently curved with the ends sticking up slightly. The prepupa dies inside the capped cell but the *capping is removed* by the workers so that they can dispose of the corpse.

The fluid around the body of the prepupa contains enormous numbers of virus particles and these get into the adult house bees when they remove the dead body. The virus accumulates in the hypopharyngeal glands and is fed to the young larvae with the brood food, so starting the cycle again. Once it is acquired, the virus remains in the hypopharyngeal glands of the adult bee and, although there is no apparent sign of disease in the affected adult bees, there are effects:

- their lives are shortened
- their development is accelerated so that they become foragers earlier than normal
- they stop feeding larvae
- very few of them collect pollen.

These changes mean that they do not infect very many larvae and the infection usually disappears spontaneously. The virus survives the winter in the glands of the adult worker bees and its effects often reappear in *spring*, when there may be a shortage of nurse bees, and then clears up. The *chewed cappings, fluid-filled sac* and *brownish colour* of the prepupa can be confused with American foul brood (AFB) and my experience is that the remains, at the right stage, can be very sticky and show a tendency to form a short rope. I once had a badly affected comb in a colony but, when I went back to the hive the next day to remove the comb and photograph it, all the diseased prepupae had disappeared. The bees had cleared them out. I never had any more problems in that colony – at least not that I saw!

What can be done about sac brood? There is no cure other than to try replacing the queen with another from a different strain as it does appear that some queens pass on genes which make their larvae more susceptible to the virus.

Black Queen Cell Virus (BQCV)

This is quite a good name as it is a virus which attacks sealed queen cells and turns them black (or dark brown). So no problem there. The dead prepupa or pupa inside the cell will have a yellow colour and a tough skin at first so that, superficially, it resembles the dead prepupa associated with sacbrood. The disease tends to attack queen cells which are being reared in numbers as a result of intentional queen rearing and the virus is interesting because it can only multiply when nosema (see page 64) is present in the colony. It does not appear to affect worker bees although it is frequently found in them.

BACTERIA AFFECTING BROOD

We come now to two troublesome and serious diseases: the foul broods.

American foul brood (AFB)

The bacterium causing this disease is *Paenibacillus larvae*. It is a rod-shaped bacterium which forms very long-lasting and tough spores. The salient points of its life cycle are as follows:

- spores are taken in by the larva in its brood food
- the spores germinate to produce vegetative cells in the gut, but these do not multiply at this stage
- the vegetative cells move into the epithelial cells lining the gut and from there into the haemolymph
- the cells multiply in the haemolymph once the larva is fully fed and sealed into its cell
- spores are formed by the vegetative cells in enormous numbers. (2500 million per individual)
- the prepupa/pupa is killed by the enormous number of cells in its haemolymph. The condition can be described as septicaemia
- cell-cleaning house bees become contaminated with the spores when they clean out affected cells.

Sealed brood affected by American foul brood (AFB)
Note the sunken cappings and the large number of empty cells

There are some fascinating points about these bacteria:

- the spores can remain infective for at least 35 years. They are resistant to heat, disinfectants and desiccation, so just leaving them to go away is not an option
- the vegetative cells are not infective
- a larva up to 24 hours old needs only 10 spores, or less, to infect it. Larvae more than two days old need millions of spores before they succumb. This is probably because the peritrophic membrane acts as a barrier to infection but takes a few days to develop in the young larva
- the bacterium can exist at very low levels without killing the colony
- once a few hundred larvae have been killed by the bacterium, death of the colony is inevitable.

What signs are there that a colony is infected with AFB?

A thorough examination of all the sealed brood is essential. There is no seasonal peak of AFB, it can flare up at any time when brood is present. To look at the sealed brood, it is necessary to shake all the bees from the comb, so, at regular intervals every beekeeper should spend a few minutes looking specifically for disease signs.

What can be seen? The bacteria only get going in the sealed larva/prepupa and usually do not kill it until after it has spun its cocoon. Following the death of its inhabitant, the cell changes in appearance: the wax capping sinks, becomes darker and may look greasy or wet. This is what you would expect when a fat, juicy insect is decaying away underneath it. The worker bees, who are far more observant than your average beekeeper, soon twig that something is amiss and chew small holes in the capping to investigate. At this stage, if the beekeeper opens the cell (s)he will find a brown slimy mess inside. When a matchstick is dipped into this and gently pulled up a brown, mucus-like 'rope' forms attached to the matchstick. This will get to about 25 mm before it breaks and is an almost foolproof test for AFB. (Be careful to destroy the matchstick.)

What happens if we do not spot the danger signals? The individual pupa continues to break down and then becomes drier, eventually, after about a month, drying up completely to form a hard, dark-brown scale which is stuck very firmly along the length of its cell and which the bees are unable to remove. At this stage the beekeeper should see more signs:

- empty cells where the queen has not laid because there are scales present. This gives a so-called 'pepper-pot' appearance

The decaying cell contents will form a mucous-like rope which is characteristic of AFB

to the brood, but other brood diseases may result in the same effect
- very dark brown scales lying along the bottom of these empty cells. These may be the only signs of the disease in dead colonies or in old equipment. They are not that easy to see if you do not know what you are looking for so it is important that all beekeepers try to receive some training in spotting them. The photograph shows how to hold a comb so that they are visible. A light source must shine into the lower part of the cell so that the scales show. A torch is a useful tool here
- there may be an unpleasant smell. This is often likened to fish glue but how many of us know what fish glue smells like? Just assume that if there is an unpleasant smell, something is wrong.

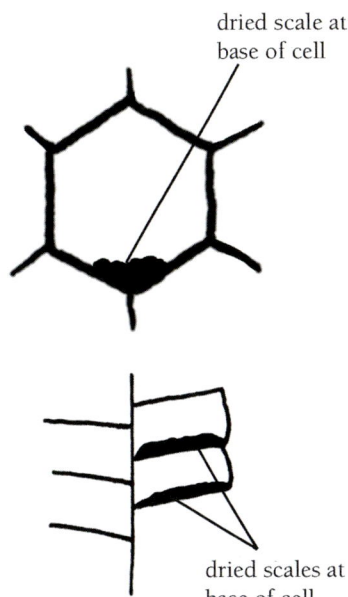

Position of AFB scales in cells

What happens to the colony? House-cleaning bees come along and try to clean up both the messy (pre)pupae and the scales, so becoming contaminated with the spores. The spores may get to every part of the colony including the honey. The house-cleaning bees soon become nurse bees, feeding young larvae, and the spores will be passed on to the young larvae in this way. The disease may be quite slow to get going, the bees may even keep it under control for a time by removing diseased larvae in the early stages, but, once established, it will romp through a colony reducing it so much that it will die. This can happen at any time of the year.

How do other colonies get infected? Remember that the spores can get to every part of a colony so they will be on, and in, combs and bees, in honey and on all hive parts. Natural methods of spread are:

- drifting, where a worker bee may simply go into the wrong hive, taking spores with it
- production of a swarm by an infected colony
- robbing. This is probably the most important bee-based method of spread. Colonies weakened, or killed, by AFB have their stores looted by other colonies. Massive numbers of spores may be carried back to the robbing colony which then becomes infected.

Frame inspection
When looking for AFB scales you must hold the frame so that the light falls on the lower wall of the cells. If there is insufficient light, use a torch

But if you want really efficient spread of the disease get a beekeeper. (S)he may:

- move combs and other hive parts from an infected to a non-infected colony

- move colonies between apiaries
- unite a weak colony (ie, a diseased one) with a strong one
- feed honey from dubious sources because it is cheap
- trap pollen from an infected colony and feed it to another healthy one
- delve into the hives of beekeeping friends wearing dirty gloves and an overall previously used for inspecting other, diseased, colonies
- hive unknown swarms near other colonies – there will be no obvious symptoms in a swarm
- buy old equipment and put bees into it without first flaming it all with a blowtorch
- move bees to areas where there are many other hives close by, ie, some migratory beekeeping.

Control of AFB

By looking at the ways of spreading the disease we can immediately see ways to control its spread.

The most important aspect of AFB control, as with any other disease, is to be observant. Look for small changes in the brood and, if in doubt, shout for help. In the UK, beekeepers are responsible for the health of their colonies and the National Bee Unit (NBU) can only function as a back-up. If a beekeeper *thinks* (s)he *may* have a hive infected by AFB (s)he *must, by law*, contact the NBU (Scottish Executive Environment and Rural Affairs Department [SEERAD] in Scotland; Department of Agriculture and Rural Development, Northern Ireland [DARDNI] in Northern Ireland), usually in the form of the local Bee Inspector, so that the bees can be checked,

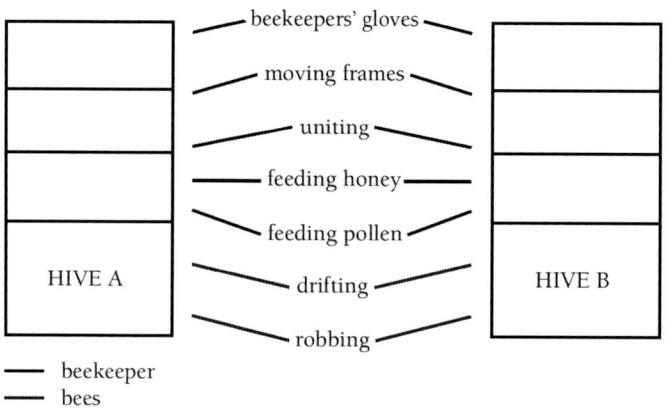

Movement of disease-causing organisms (pathogens) between hives

because AFB is a *notifiable disease* (under the Bee Diseases and Pests Control Order 2006). The problem will often be due to another, less serious, condition, but it is always better to be safe than sorry.

The Bee Inspector will inspect any suspect colony and ascertain whether AFB is present or not. This may involve visual diagnosis, use of a lateral flow device (like a pregnancy testing kit) or laboratory inspection of the comb. Once diagnosed, the whole system for dealing with a notifiable disease comes into play: the colony affected by AFB is destroyed by the Bee Inspector. The bees are killed, all bees and combs from the affected hive are burnt and the remaining parts of the hive are thoroughly sterilised with a blowtorch. Any other equipment, such as hive tools, which has been in contact with the colony, is washed in hot water containing washing soda. All fabrics such as overalls are washed in hot water. A *Standstill Order* is placed on the apiary, as soon as the disease is suspected, and this stays in place for at least six weeks after destruction of the colony when, following a further inspection and if all is clear, it is lifted. During this time no bees or equipment may be moved out of the apiary.

The incidence of AFB has been considerably reduced over the years and is kept at a very low level in the UK due to this destruction policy and any other procedure is illegal. In some other countries there has been routine, and often prophylactic, use of antibiotics to keep AFB under control and many problems are now being experienced because the bacterium has developed resistance to the antibiotics. The antibiotic treatment does not kill spores so AFB is breaking out in many hives.

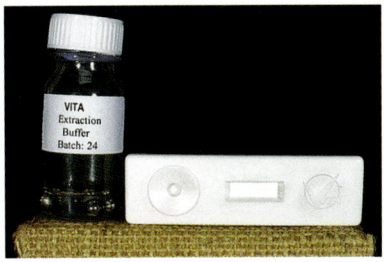

It is possible to diagnose AFB with a lateral flow device. The suspect material is macerated in the buffer solution from the bottle and three drops of the liquid are put in the circular window. One blue line indicates a negative result, two blue lines indicate that AFB is present. The result should be apparent in approximately three minutes

European Foul Brood (EFB)

When we talk about American foul brood we can see that there is an inevitability about its progress in a colony. Now we are going to look at European foul brood, (EFB), which is a complete contrast. It always seems a pity that the two diseases have similar names and are often considered together but that is an historical quirk and there is nothing we can do about it.

European foul brood is an enigma in many ways. In the individual colony or apiary it appears to come and go, in many parts of the world it is considered a comparatively minor problem, but it can, and does, destroy many colonies. So what causes it, how is it spread and what can be done about it?

The bacterium responsible for causing the symptoms of EFB is *Melissococcus plutonius* (formerly *Melissococcus pluton*, before

Unsealed larvae showing the characteristic discolouration, twisting and waxy appearance associated with European foul brood (EFB)

that *Streptococcus pluton*, and before that *Bacillus pluton*). Life can become quite confusing, however, because, unlike AFB, a larva infected with the **M. plutonius** bacterium, becomes infected with other bacteria which move in as secondary infections. The main ones are:

- *Paenibacillus alvei*
- *Brevibacillus laterosporus*
- *Lactobacillus eurydice*
- **Enterococcus faecalis**.

There has always been confusion about these different bacteria, their relationship with **M. plutonius** and their effects on the larvae. We now know that **M. plutonius** is not such a simple organism but is quite variable, existing in a number of different forms, some of them common, some very rare and isolated, some virulent, some less so. EFB becomes more of an enigma the more we know about it.

How does the disease affect the larva?

Having introduced confusion and uncertainty at the start, we will go on to the development of the disease in the individual larva. **M. plutonius** infects the very young larva and multiplies in the ventriculus (true stomach). It never moves anywhere else in the body of the larva and, in a five-day old infected larva, the ventriculus will be occupied by many bacteria. The effect of the bacteria is to starve the larva by consuming the food in its stomach. An infected larva showing the signs of EFB will:

- lie uncomfortably in its cell, twisted and misshapen
- be a funny colour, off-white, greenish or brown
- lose its segmentation so that it looks like a bit of melted wax.

When a larva is infected one of three things may happen:

- the nurse bees will detect the infection and throw the larva out of the colony
- the larva will die because the bacteria have consumed its food. It usually dies before its cell is sealed and will dry to a rubbery scale if left in the cell
- if the larva is very well fed, so that there is enough food for larva and bacteria, it will survive and pupate.

The outcome will depend on the hygienic behaviour of the house bees and the proportion of nurse bees to unsealed brood:

- hygienic bees will detect and remove diseased larvae more readily
- a high proportion of nurse bees to larvae will provide lots of brood food and the larva will survive, although it will be weakened and will spin a poor cocoon.

How does it spread in the colony?

There are three important facts involved in the spread of *M. plutonius* in a colony:

- *M. plutonius* never forms spores. The normal vegetative cells are infective and produced in huge numbers in the infected larva
- the contents of the ventriculus of a larva, and so the bacteria, are 'sealed in' until the larva pupates and the connection between the ventriculus and the hindgut opens. Then all the waste and bacteria, which have been stored in the gut during larval life, pass out into the cell.
- very young adult bees clean out cells and later produce brood food and feed young larvae.

If we take these together we can see how the disease spreads through the colony. Infected larvae which survive to pupate discharge their gut contents into the cell. The cell is then cleaned by the house bees which pick up the bacteria and, subsequently, feed them to young larvae in brood food. Where a weakened larva has spun an inadequate cocoon, the bacteria will be more accessible to the house bees.

A fluctuating problem

EFB can be a puzzle because of the way it appears to come and go. Often colonies can survive with it for many years before it gets on top of them. This is because of the activity of the young bees in removing diseased larvae at an early stage and because the bacteria do not really start to kill large numbers of larvae unless those larvae are on short rations. So, very early in the year there may be an abundance of nurse bees, the larvae will survive and we will not notice anything wrong, although the foundations are being laid for later trouble.

Once the first nectar flow starts, young bees are recruited to foraging duties, there are fewer nurse bees to feed the brood, which is rapidly expanding, larvae are readily infected from the cell contents of previously infected larvae and disease symptoms appear. Left untreated, this may clear up when the nectar flow ceases and an abundance of nurse bees builds again but if it has become severe, the colony may become very badly affected and die. The disease will be more troublesome where the season is a stop/go affair with good honey flows interspersed with periods of dearth.

How does it spread to neighbouring colonies?

In infected colonies, bacteria will be on the comb and other parts of the hive. So, if you handle these and then manipulate another colony, or visit a friend's apiary, you are very likely to take the bacteria with you. Moving infected combs about from hive to hive is a very efficient way of spreading disease organisms around. Natural means of spread by the bees occurs when there is robbing or drifting, or the colony swarms.

What do we do about it?

The first thing I have to say is that EFB is a notifiable disease in the UK (under the Bee Diseases and Pests Control Order 2006) at the present time so, if you suspect you *might* have it in one of your colonies you **must** notify the NBU or other appropriate authority, as for AFB. Diagnosis may be visual (often difficult), by lateral flow device or by microscopic inspection of the bacteria present. A Standstill Order will be issued and there will then be a choice of treatments:

- destroy the colony by burning if it is severely infected
- treat a light infestation with the antibiotic oxytetracycline (OTC) and then inspect again after eight weeks. The OTC is dissolved in sugar syrup and sprinkled directly onto the brood nest. It must be administered by a bee inspector
- shake the bees onto clean foundation in a new box (this is called a shook swarm), in addition to treating with OTC, and destroy the brood. This method of treatment is showing good results. It may be possible to obtain good levels of control using the shook swarm method without the additional use of OTC, so reducing the use of antibiotics.

Many workers in the field of EFB believe that the organism responsible is endemic in hives. Work done some years ago seemed to suggest that this is so and that the actual disease results from a combination of factors including stress. Assuming that this is the case, as beekeepers we can apply the usual preventative measures:

- provide the colonies with new foundation and clean boxes regularly
- fumigate all out-of-use brood boxes and brood frames with 80% ethanoic (acetic) acid
- ensure that our beekeeping clothing is clean and use disposable gloves
- use extreme caution when transferring frames from one colony to another, particularly if they contain brood, and when uniting colonies
- never use second-hand comb and singe any second-hand brood boxes with a blow torch
- inspect regularly for suspicious larvae. Sometimes larvae die after sealing and these may be confused with those affected by AFB but any abnormality should be investigated.

FUNGI AFFECTING BROOD

There are only really two brood diseases in this section. One of these, *chalkbrood*, is a very important and widespread condition and the other, *stonebrood*, is one of those which appear in all the books but nobody ever seems to have seen – or at least, recognized.

Chalkbrood

This is such a common disorder that every beekeeper comes across it sooner or later. It can cause the death of many larvae and prepupae and is caused by a fungus which delights in the name of *Ascosphaera apis*. The larva takes in the spores of the fungus with its food and, once inside the gut, the spores start to grow. They produce hyphae which are like fine white cotton threads and characteristic of fungi in general. These grow through the gut wall, continuing to spread through the body of the larva and eventually grow out through the cuticle. The mass of intertwined hyphae is called a mycelium and eventually the larva becomes a swollen mass of fluffy white fungus with a small yellow lump where its head used to be.

As you might expect, the larva dies, but often not until after it is capped. The worker bees, which can detect the dead larvae, uncap and remove them.

By this time the remains of an infected larva will have dried to

Chalkbrood is caused by the fungus *Ascosphaera apis*
The head remains as a yellow structure

a hard lump called a *mummy* (as in the Egyptian, preserved type, not the nurturing kind) which may be white, grey or black. The dark-coloured ones are producing spores which are shed in huge numbers into the hive. These are sticky, very resistant and can remain infective for at least 15 years.

Where there are many affected larvae, the mummies will be found on the hive floor or in front of the hive entrance. In these circumstances there will be many empty cells among the capped brood and a 'pepper-pot' appearance will result.

We can summarise the signs of the disease:

- white, fluffy fully-grown larvae or prepupae with tiny yellow 'heads' which can be seen when cappings are removed. As the bees remove the cappings to get at the dead brood, the larva/prepupae are often visible in open cells, sometimes in large numbers
- hard 'mummies' on the floor of the hive or in front of the entrance. There may be large numbers of these
- pepper-pot appearance to the brood.

Chalkbrood 'mummies' can be found on the floor of the hive or outside it where infection is severe

When are you most likely to see chalkbrood?

It often flares up in the spring when the colonies are expanding rapidly. There are not sufficient nurse bees to keep all the brood

WHEN THE KIDS ARE ILL

warm and the slight chilling that results, while not sufficient to kill the larvae, seems to favour the growth of the fungus.

Small colonies are susceptible because they cannot retain heat as well as large ones and, as usual, the beekeeper is the biggest threat. We divide colonies, give brood to colonies which cannot cover it, make up queen mating nuclei in little boxes and expose the brood to cooling air while we inspect it.

What can you do about it?

Not a lot! Avoid chilling the brood. Regularly replace comb where colonies have suffered from chalkbrood. Some authorities maintain that some strains of bee may be more likely to develop chalkbrood than others. If this is so, and it is open to dispute, the disease may be reduced by requeening with a queen from a different strain.

As with all matters pertaining to bee health, it pays to have young queens in strong healthy colonies. Such colonies are more likely to get on top of any infection.

Pepper-pot brood (with many scattered empty cells) can indicate various brood conditions including chalkbrood and AFB

Stonebrood

Although this disease is very uncommon, it is of interest mainly because the fungi responsible for it can not only infect other species, particularly birds, but can also cause breathing problems in humans. It is similar to chalkbrood in the initial stages and the fungus follows the same pattern of infection, although it can also attack the larva from the cuticle inwards. Again the larvae become white and fluffy but do not usually die until they have been capped. Then they become very hard, hence the name (chalkbrood mummies can easily be broken with the fingers). The colour is different too: the dead prepupae become yellow and then green when spores form. In some cases adult bees can be affected, with their abdomens becoming hard. There are several fungi which can cause stonebrood, among them *Aspergillus flavus* and *Aspergillus fumigatus*, common soil- and air-borne fungi which are opportunistic pathogens on plants and some animals.

Humans, usually those with compromised immune systems, may be affected by these fungi and they can cause serious disease. Care must be taken when handling or destroying combs infected with stonebrood.

Pupae killed by stonebrood
This disease is rarely seen

5 PROBLEMS WITH THE GROWN-UPS

Diseases of adult bees often go undetected because there is no outward sign in the individual, other than that it is dead. Usually the beekeeper is unaware that the individual bee is dying early, but the overall effect is a failure of the colony to build up or the gradual reduction in the size of the colony culminating, in some cases, in its complete collapse. In looking at the adult bee diseases, we will follow the same pattern as with the brood diseases, so we will start with viruses.

VIRUSES CAUSING ADULT BEE DISEASES

The problem with adult bee viruses is that, because we cannot see them, cannot treat them and the affected bees may have no visible symptoms, we may ignore them, but, once rampant, they all result in the early death of the individual bee and, when sufficient bees are affected, the colony may collapse. Some are associated with other diseases and probably increase the severity of such diseases, and others are spread and activated by *Varroa destructor* and, possibly, other mites. I have listed the main ones and have tried to sort them into some logical system:

- **A** **The cause of a huge argument!**
  Chronic bee paralysis virus

- **B** **Associated with *Varroa destructor***
  Slow bee paralysis virus
  Deformed wing virus
  Varroa destructor virus 1
  Kakugo virus
  Acute bee paralysis virus
  Israeli acute paralysis virus

Kashmir bee virus
Cloudy wing virus

    C   **Associated with *Nosema Apis***
Filamentous Virus
Bee Virus Y

    D   **Associated with *Malpighamoeba mellificae***
Bee Virus X

Chronic bee paralysis virus (CBPV)

The condition caused by this virus is often referred to as 'paralysis' but we must be pedantic, so that it is not confused with the other two types of paralysis.

Chronic bee paralysis virus was discovered in 1961 beginning a long-running debate, principally between the late Brother Adam, the renowned bee breeder from Buckfast Abbey, and the late Leslie Bailey, a leading worker on honey bee viruses at Rothamsted Experimental Station (now Rothamsted Research), concerning the cause of the 'Isle of Wight Disease' which destroyed many colonies in Britain in the early part of the twentieth century. The blame for this epidemic was originally laid at the door of *Acarapis woodi*, a mite which causes acarine, and Brother Adam adhered to this belief. However, the viruses were not known then and subsequent work by Leslie Bailey showed that CBPV was present in bees showing symptoms associated with Isle of Wight Disease. The argument raged on and it is most likely that the huge numbers of colony deaths which occurred between 1904 and 1920 were due to a variety of reasons, involving more than one condition. We will never know for certain.

But back to CBPV. What does it do to the individual bee? There are two quite distinct sets of symptoms:

Type 1 syndrome

- bees with trembling wings and bodies.
- inability to fly with sometimes most of the colony leaving the hive and crawling about on the ground outside.
- bloated abdomens due to a build up of fluid in the honey stomach.
- K-wings, where the wings are partially spread and form the letter K, with the body being the upright part.

- onset of dysentery due to the fluid in the body.
- deaths of the individuals often leading to collapse of the colony, sometimes very rapidly.

Type 2 syndrome

- bees appear black and shiny, and are hairless.
- bees have a broad abdomen.
- the affected bees are nibbled by other bees (which is why they are hairless) and refused access to the colony, which makes them appear like robbers as they try to get back home.
- after a few days the bees tremble, are unable to fly, and die.

These two syndromes may occur in one hive, but usually one is predominant.

I have, over a period of years, lost colonies showing symptoms of the Type 1 syndrome. They had heavy acarine infestations but I have no way of knowing whether they had CBPV as well. I am not aware that there has ever been any suggestion that the two conditions are linked, ie, that the acarine mite can act as a vector for CBPV or can interact with it, but who knows what awaits discovery?

Viruses associated with *Varroa destructor*

Some of these viruses were just text book jobs before varroa arrived, then, suddenly, they became of concern to all beekeepers. They are excellent examples of how viruses can remain dormant inside an organism until the right conditions occur, when they suddenly erupt.

Colonies infected with *slow bee paralysis virus (SBPV)* may collapse late in the year even after most of the mites have been destroyed. This virus infects larvae as well as adults and, once rampant in a colony, carries on multiplying and spreading even if there are no longer any mites around. In the early days of varroa, this virus was the cause of many doubts about the efficacy of the varroa treatments available because beekeepers who had treated their colonies, and killed many mites, still lost their colonies later in the year. This was usually because they had treated too late and failed to protect the winter bees.

We have now all seen bees with little deformed wings and stunted bodies (in other people's hives of course). Such bees are

These two bees are showing deformed wings, probably caused by deformed wing virus

infected with *deformed wing virus (DWV)* and soon die. What is not always realised is that many bees that have deformed wing virus do not have deformed wings.

Varroa destructor virus-1 (VDV-1) and *Kakugo virus (KV)* are viruses closely related to DWV. They are all able to transfer genetic material between each other so that a complex of viruses builds up. They are also able to multiply in the varroa mite. Early work on KV showed an association with the presence of the virus in the bees' brains linked to extremely aggressive behaviour.

Acute Bee Paralysis Virus (ABPV), *Kashmir Bee Virus (KBV)* and *Israeli Acute Bee Paralysis Virus (IAPV)* form another group closely related to each other. ABPV is activated by varroa mites and then spread around by them. Affected adult bees may 'feed' larvae with the virus so that they become infected. The combination of KBV with varroa leads to the very rapid decline of a colony, infected bees dying within a few days. IAPV has been associated with the deaths of many colonies in the USA from Colony Collapse Disorder (CCD). It appears not to be the main cause of such losses but may contribute or serve as a marker for the condition.

Cloudy wing virus (CWV) is at the end of this section because it is by no means certain that it is associated with varroa, but it probably is. It causes the wings to lose their transparency and the bee soon dies.

All of these viruses cause the death of the individual bee but you do not know that they are present in your colonies until the symptoms appear and by then it is probably too late to save the colony. So you must assume that some of these viruses are present in your bees and ensure that you keep the number of mites as low as possible to minimise the effects.

Undoubtedly, as more research is undertaken into the whole problem of viruses and varroa and use is made of ever more sophisticated techniques for analysing genetic material, the whole subject will become even more complex. We must not forget that viruses are continually mutating and evolving so that the virus landscape will continue to change.

Viruses associated with *Nosema apis*

By 'associated with' in this instance we mean that these viruses only multiply when *Nosema apis* is present. *Bee virus Y* apparently causes no symptoms of its own but *filamentous virus* may cause the haemolymph (blood) to become milky in appearance. It is very likely that these viruses worsen an infection by *N. apis*. Both of them probably get into the bee through its gut, so the fact that Nosema weakens the gut wall will allow them easier entry than they would get through the wall of a healthy gut.

Virus associated with *Malpighamoeba mellificae*

Bee virus X is not totally dependent on the presence of *M. mellificae* (the cause of Amoeba) but seems to appear more frequently when it is present. Bees infected with the virus have much shorter lives and it is often associated with deaths of colonies in spring.

There are a few other adult bee viruses which seem to be of academic interest at the moment and there are probably others which we do not know about so this account is not complete. Undoubtedly viruses as a whole are the cause of many colony losses even if the beekeeper concerned does not always recognise the fact. Attempts are being made to develop treatments for viruses. These include interfering with the genetic message which the virus passes to its host cell, causing that cell to produce virions. This is done using RNAi (interference RNA).

DISEASES CAUSED BY MICROSPORIDIA

Microsporidia are single-celled organisms which form spores and generally affect the gut. As we saw in Chapter 3, they have a unique

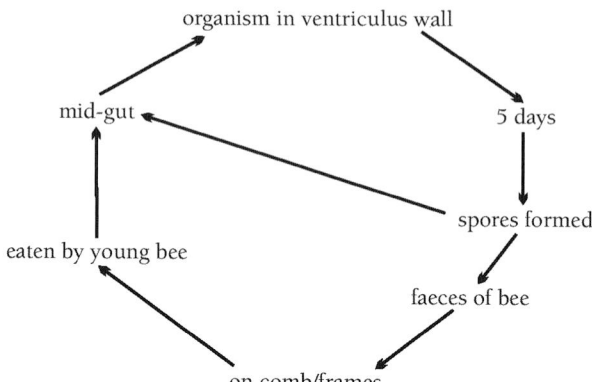

The life cycle of *Nosema apis*

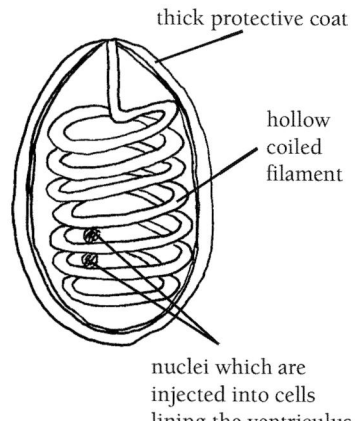

A *Nosema apis* spore
(*Based on a diagram in* Honey Bee Pathology *by Bailey & Ball*)

way of getting into the cell of another organism (the host) by 'firing' a special hollow tube into the targeted cell. The microsporidian cell contents pass through the tube and into the host cell, where they begin multiplying. We are concerned with two species:

- *Nosema apis* – a familiar pathogen.
- *Nosema ceranae* which was found on honey bees in Europe, in Spain, as recently as 2005. It did not even have a name until 1996, when it was discovered in *A. cerana* by Ingemar Fries. It has spread worldwide in the past 20 years or so.

Incidentally, other species of Nosema affect many different insects (and other animals) and one of these, *N. bombycis*, caused very heavy losses in the silkworm industry in the late nineteenth century until Louis Pasteur discovered its cause and a strategy for dealing with it.

The life and spread of *Nosema apis*

The organism gets into a bee through its mouth. (If you're reading this over a meal, skip the next bit until you've finished eating!) Faeces of infected bees contain millions of spores of *N. apis*. Some of these are found on the comb and frames and the only way a house bee can clean up the mess is to eat it. The spores then pass into its mid-gut (ventriculus) where they germinate and infect the cells lining the mid-gut. These are called epithelial cells and, once inside the cells, the invading *N. apis* multiplies, feeding on the cell contents and finally killing the cell. New spores are formed and, under ideal conditions, this can happen in five days. The host cell

PROBLEMS WITH THE GROWN-UPS

breaks down and the spores are released into the bee's gut, which may contain between 30 and 50 million spores when the infection is fully developed. Some of these can invade new host cells and the others pass out with the faeces, starting the whole cycle again.

Spread can also be from water sources containing bee faeces.

What does it do?

- there is no way of telling an individual bee has nosema by looking at it.
- in spring and summer, the lifespan of an infected bee is halved.
- the hypopharyngeal glands of the infected bees do not develop fully so that they are unable to feed brood. The bee tends to become a guard or forager much earlier in its life than is normal.
- internally there are changes within the epithelial cells of the mid-gut and their normal function, producing digestive enzymes, becomes disrupted and may cease altogether.
- in winter bees, there is less protein stored in the fat bodies of infected bees and there are more amino acids circulating in the haemolymph.
- also in winter, rectal contents increase mainly due to an increase in water.
- the increased amount of waste in the rectum may be voided in the hive and around the entrance, a condition known as *dysentery*.
- queens which become infected (fairly unusual) stop laying and die or are superseded.
- there is a reduction in honey yield in colonies suffering from the disease.

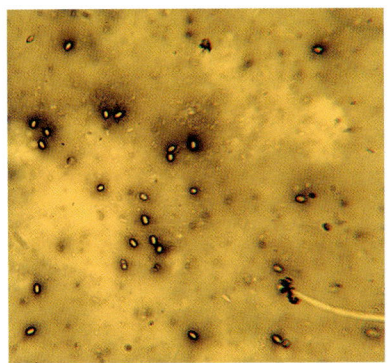

***Nosema apis* spores stained with nigrosin**
(magnification x400)

Diagnosis

Because there are no visible symptoms in the individual bee, a high power microscope, with a magnification of about x400, is needed and then it is very easy. Thirty bees are needed and their abdomens are crushed in 10 ml of water. A drop of the resultant mess is put onto a microscope slide and spread into a smear. It may be stained with nigrosin or left unstained. The spores of *N. apis*, if present, will look rather like tiny rice grains and the more there are the higher the level of infection. It is possible to estimate the amount of nosema in a colony by using a simple table but to do this a graticule (a method of measuring under the microscope) is needed on the microscope. The percentage infection is calculated by counting the number of spores in a given area. Using this method, 28 bees are needed to give a 95% chance of finding an infection affecting 10% of the bees or a 99% chance of discovering an infection affecting 15% of the bees. From these figures it can be seen that 30 bees will provide an indication of a comparatively low level of infection. Of course, if the infection is much worse, theoretically fewer bees are needed and, where 70% of the bees are infected, only four bees need to be used to give a 99% chance of finding the spores. It is all a matter of statistics.

Cyclical nature of the disease

Nosema only appears to cause severe problems in winter. A colony which is infected in the autumn may very well fail to survive the winter or early spring. It is one of the causes of, so-called, spring dwindle, where the colony fails to build up and finally dies. A colony which has nosema will lose more bees during the winter period (since they don't live as long and their fat bodies are not so well supplied with protein). There will also be a problem with the reduction in the number of bees which will be able to act as nurse bees in the crucial early part of the year (because the hypopharyngeal glands do not develop properly). Where dysentery is severe, more bees will become infected as they clean up.

Summer infection is less serious and tends to clear up. This is because, in summer, a bee is always able to empty its rectum outside the hive and so the infection is removed from the colony.

So far, the comments have been general and apply particularly to *N. apis* so now we have to muddy the waters somewhat.

How does Nosema caused by *N. ceranae* differ?

- *N ceranae* in Spain proved to be very virulent, much more so than *N. apis*. However, this has not been shown elsewhere. This may be due to different strains (haplotypes) of the pathogen or to different subspecies of bees reacting differently or to the fact that Spain is a hot country.
- the spores of *N. ceranae* are smaller and more slender than those of *N. apis*, but to most observers it is impossible to tell the difference between them under the microscope.
- *N. ceranae* does not produce so many spores as *N. apis*
- *N. ceranae* attacks the basal cells of the gut wall as well as

PROBLEMS WITH THE GROWN-UPS

the epithelium and this damage provides an entry point for viruses and bacteria. This may be the reason that *N. ceranae* may depress the immune system of the bee whereas *N. apis* does not seem to do so.
- there is no seasonal variation in *N. ceranae*.
- dysentery does not occur in colonies affected by *N. ceranae*.
- *N. ceranae* is apparently spread in the pollen as it is moistened with nectar from the bee's crop. In this way it can infect very young bees and this may account for its non-seasonal variation.
- *N. apis* spores are damaged by heat but not by freezing, whereas *N. ceranae* spores cannot withstand freezing but are unaffected by heat.
- colonies affected by *N. ceranae* alone, or a combination of the two species, are more likely to collapse than those affected by *N. apis* alone. This is probably due to the seasonal nature of *N. apis* which usually clears up in the summer.

This is a brief summary of the features of *N. ceranae* and there is clearly much more to be learned about it in the coming years.

Complications

There always are! There are three viruses which multiply only in the presence of Nosema. We have already discussed two of them (see page 63), namely *filamentous virus* and *bee virus Y*. In addition, we must mention **black queen cell virus (BQCV)**, an aptly named virus since it causes the deaths of queen pupae and the cells containing them become dark brown or black. The queen larvae are infected with *N. apis* as worker bees feed them. The individual eventually dies in the prepupal or pupal stage and is at first yellow, the colour later changing to brown.

The other condition that interacts with Nosema is dysentery. (Note that dysentery is not a disease and does not appear to be induced by *N. ceranae*.) Nosema does not directly cause dysentery but dysentery always makes Nosema much worse. As we have seen, *N. apis* may cause an increase in the water content of the rectum. Where bees are confined within the hive, this increase in the rectal contents may cause the bee to defecate in the hive, so leading to more bees being infected when they clear up the mess. This is the major reason why an autumn infection can spell disaster for a colony.

The final complication, as in many other diseases, is *stress*, particularly any factor which keeps the bees confined in their hive for a lengthy period of time. So, migratory beekeeping, poor

Brown stains on the front of the hive are due to dysentery

weather and too many hives in an area can all lead to an increase in the incidence of nosema. Particularly damaging is a poor pollen supply and lack of variety in pollens and this can arise, not only where bees are on a monoculture for long periods and where they do not have access to a range of different pollens, but also where the weather is bad for a prolonged period of time.

Control

Clean combs are essential. Get the colony onto clean comb and remove the old ones. The late Leslie Bailey devised a system called the *Bailey comb change* with the primary aim of reducing nosema, but it works equally well for other disease organisms carried on the comb. The process has to be timed with a nectar flow, or the colony must be fed because they will have a great deal of foundation to draw out. The diagram illustrates the method. If the hive has supers on it, they are put above the upper brood chamber over a second queen excluder. The two brood chambers are left in place until all the brood in the lower box has emerged and then that is removed. The single frame in the new brood chamber is removed at a later date.

Old combs can be cleaned using ethanoic (acetic) acid (see page 35). This will kill the spores of Nosema and many other disease-causing organisms. However, it is usually better just to burn them and scrub the frames in very hot soda solution.

The usual advice of keeping colonies strong with young queens is sound and keeping them in the sun, at least for the autumn/winter/spring period, can also help. Opening colonies during cold spells of weather can make nosema worse.

There is no chemical treatment available for Nosema in the UK, but various hive cleansers and tonics are on the market and may help.

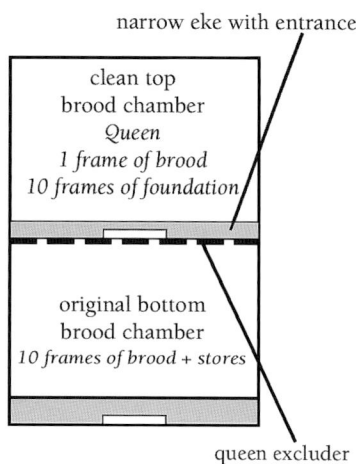

Bailey comb change

A DISEASE CAUSED BY A PROTOZOAN

Protozoa are microscopic single-celled (or colonial) animals. They are found everywhere and some of them cause disease (a human example is the protozoan causing malaria). Most of us heard about them when we learned about *Amoeba* at school. Their reproduction is usually very simple – they simply split into two.

In the honey bee, we are interested in an animal which goes by the complicated name of Malpighamoeba mellificae. Although this

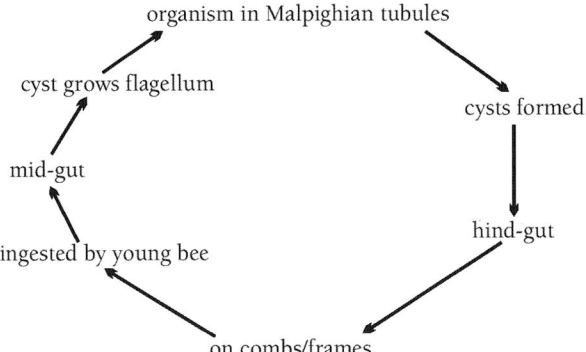

Life cycle of *Malpighamoeba mellificae*

is a long name it tells us a lot about the animal: that it affects the Malpighian tubules, that it is an amoeba-like animal, and that it belongs to the honey bee (*A. mellifica* was the old-fashioned name for the honey bee). It is commonly referred to, by beekeepers, simply as Amoeba.

How does *M. mellificae* live?

The little animal is able to produce a protective wall around itself to form a cyst and, in this form, it is taken into the adult bee. Inside the gut, a mobile amoeba, complete with flagellum, emerges from the cyst and moves into the Malpighian tubules through the openings which connect them to the gut. Once there, the amoeba is able to attack the lining cells of the tubule and, after three to four weeks, form new cysts which pass out of the gut with the faeces.

What does it do?

This is the problem with amoeba. It does not appear to have any effect on the infected bee. There are about one hundred Malpighian tubules so presumably the bee can survive quite happily with one or two out of commission. *M. mellificae* is often found in bees also infected with *N. apis* and it would be reasonable to suspect that it could aggravate that condition. Viruses may also be involved. Its reproductive rate is much slower than that of *N. apis* but both organisms are spread in the same way. There is generally a peak of infection in early summer and the condition then clears up.

Microscopic examination of a sample of bees, using exactly the same method as for *N. apis*, is the only way of determining whether the disease is present or not.

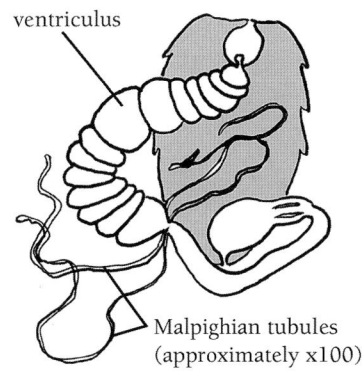

Dorsal view of the digestive system

Nosema apis *multiplies in the cells lining the ventriculus and* Malpighamoeba mellificae *attacks the cells in the thin walls of the Malphighian tubules*

CHAPTER 5

What to do about it

Clean comb is the best protection against *M. mellificae* since the cysts remain on the combs for some time, ready to infect other bees.

DISEASES CAUSED BY MITES

Mites can be extremely troublesome little animals. Many live blameless lives in the soil and in many other habitats, but others attack animals and plants. They belong to that huge Phylum, the Arthropoda. This is the biggest Phylum in the animal kingdom with 13 Classes and includes millipedes and centipedes, insects, king crabs, spiders and their relatives, and crustaceans. The classification of the group is the subject of much debate and disagreement amongst biologists who worry about such things but the mites belong to the group called the Arachnida which includes spiders, harvestmen, mites and ticks.

There are two mites which we need to look at in detail. One of these is *Varroa destructor* which is going to have a whole chapter to itself and the other is *Acarapis woodi* which we will consider now.

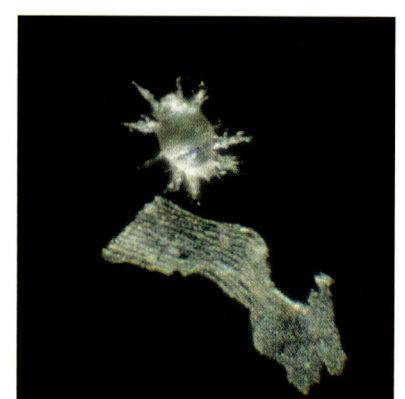

An acarine mite outside the trachea

More questions than answers – Acarine

As we saw on page 60, *Acarapis woodi* is one half of an historical argument which reaches back to 1904, when colonies of honey bees on the Isle of Wight began to die in large numbers from unknown causes. The condition, called Isle of Wight Disease, spread rapidly and was attributed to *A. woodi* in 1921. Eventually the problem diminished, there were huge imports of bees to help get beekeepers back on their feet, and the whole thing became a footnote in the history of beekeeping. It was said that the colonies of British bees which did not succumb to the disease had developed immunity to it. Some 50 years after the Isle of Wight outbreak, the discovery of chronic bee paralysis virus (see page 60) opened up a heated argument about the real cause of the original problem. The mite is now present in most parts of the world, including the Americas, Europe, Asia and parts of Africa. It often goes by the alternative name of 'tracheal mite'. It is still, apparently, absent from the countries of the Pacific including Australia and New Zealand.

The mite

- belongs to the family Tarsonemidae.
- the adult female is about 100 µm wide by 175 µm long

(depending on whether you include the hairy bits or not) and the male is smaller.
- usually lives inside the large tracheae of the honey bee. These lead from the first pair of spiracles situated between the prothorax and mesothorax and are the main route for incoming air. Mites have also been found in air sacs in the head and abdomen.
- feeds by piercing the cuticle inside the tracheae and sucking the haemolymph.
- each female mite lays between five and seven eggs.
- eggs hatch into nymphs between three and six days after being laid.
- adult females develop about 14 days after the eggs are laid, males a few days earlier (depending on which book you read).
- the mite then begins to resemble a hitch-hiker. Each one leaves its trachea, crawls up a hair and (this bit I like) hangs on by one or two hind legs while it waves the other legs in the air and waits for a suitable young bee (less than nine days old) to pass by. It then grabs hold of its hairs and crawls down to the first thoracic spiracle, drawn by the vibrations of the wings and the puffs of air.

Diagnosis

To confirm the presence of acarine, a dissecting microscope or powerful hand lens, with a magnification of at least x20, is needed.

A sample of 30 flying bees (for the same statistical reasons as with Nosema) must be taken and killed, then each one pinned, through the middle part of the thorax, to a sloping piece of cork. Either a pair of pins or a double needle is required, otherwise the bee will just whiz round when you attempt to dissect it.

The head and front pair of legs are removed by pushing them off gently with a pair of forceps. So far, so good, but the next bit is the dodgy part. There is a narrow 'collar' around the thorax, which is visible once the head has gone. It is actually the cuticle of the prothorax. This has to be removed with a pair of forceps and then the pair of large tracheae in the mesothorax can be seen clearly.

If there is brown or black coloration in one or both of these, this indicates the presence of *A. woodi*. If more than just the collar is removed accidentally, the tracheae usually disappear completely and you have to start with a fresh bee. The bees need to be freshly killed for this dissection.

A bee pinned ready for an acarine dissection

First stage in an acarine dissection, collar intact

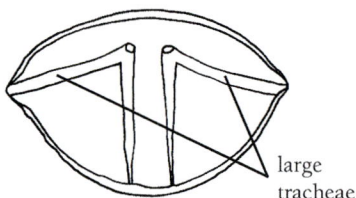

Looking for acarine in the large tracheae of the thorax
If mites are present there are dark brown/black areas in the tracheae

Final stage in an acarine dissection (right)
Part of the trachea on the left has broken but the other can be seen clearly. This bee is healthy

Recent events

It is a few years now since I started taking more than an academic interest in acarine. Then I saw one of my colonies, and the colony of a friend, die following the exodus from the hive of many crawling bees with apparently dislocated wings. Dissection (done slowly and laboriously because I'm not very good at it) showed blackened tracheae in about 80% of the bees in each case. I was able to dissect out large numbers of adult mites from the tracheae. This happened in February and I went on to lose a few more colonies, never more than two at a time, over the next few years, with the same symptoms. These, too, succumbed in February and were heavily infested with acarine. I heard of quite a few more colonies dying with the same symptoms but unfortunately no dissections had been done.

Some eminent research workers in this country, with enormous experience and knowledge, are very sceptical about acarine as a major cause of disease and the only facts available seem to be that it shortens the life of the individual bee and, where infestations are very heavy (more than 30% of bees in the colony) during the winter, severe spring dwindling can result in the death of a colony. The older bee books give the signs of acarine as large numbers of bees crawling about in front of the hive on the ground and on plants. These bees are unable to fly and their wings are held at funny angles, with the forewings often folded over the back and the hind wings sticking out. This is described as *K-wings*. It describes exactly the symptoms I have seen in my colonies. However, there is no proof that these signs are due to *A. woodi* and no-one has yet proved that acarine, on its own, is a direct threat to a colony.

The American experience

Over the last few years American beekeepers have become increasingly concerned by tracheal mite and, in 2000, Professor Mark Winston gave a lecture which set me thinking again (another slow and laborious process). He detailed an experiment where three groups of colonies had been monitored for two years. One group was infested by varroa, a second group was infested with *A. woodi* and the third group was infested with both. By the end of the first year of the experiment, 80% of the colonies in the third group was dead, while groups 1 and 2 all survived. His opinion was that acarine on its own is not really a problem but, combined with varroa, it becomes lethal. His recommendation was that all colonies should be regularly monitored for acarine and varroa.

K-wings – acarine or chronic bee paralysis virus (CBPV)?

Unanswered questions

- is *A. woodi* a serious pathogen or merely an irritant, shortening the life of the bee a little?
- is acarine directly associated with any viruses?
- are the signs, seen at many hives, of crawling bees, symptomatic of acarine or due to something else?
- are the deaths of colonies with high levels of acarine due to that or to other things, possibly a combination of pathogens?
- does a combination of varroa and acarine inevitably spell disaster for a colony?

I do not know the answers to these questions. Does anyone? The controversy looks set to continue.

Control

There is no chemical treatment for the control of Acarine but the chemicals used for varroa control seem to have reduced it significantly. Certainly thymol appears to be effective and its widespread use has undoubtedly reduced acarine to a very low level.

6 MIGHTY MITES

This is probably the most important chapter in this book because it is all about a tiny creature which has caused more changes and problems to honey bees and beekeepers worldwide than anything else in the past hundred years. The creature in question is *Varroa destructor*, the varroa mite. I am going to concentrate on the life of varroa first of all because, as all great generals will tell you, it is only by understanding our enemies that we can hope to defeat them.

In Chapter 3 we looked at the place of *V. destructor* in the great scheme of things and we now know that there are at least four different species of varroa mite recognised, of which *V. destructor* is one. To confuse matters further, there are several forms of this mite but the one which has caused such havoc throughout Continental Europe and Britain is the Korean type. All these facts are of little relevance when we are battling to control varroa in our colonies, but they are interesting.

The mite is found in almost all parts of the world where honey bees are present. It was first found in the UK in 1992, with the first sighting in Devon, but it was rapidly found in other places in the south and it had clearly been here for a few years. It then started its spread throughout the country, aided by its allies, beekeepers. So how does it live, what does it do precisely and why has it caused such devastation among honey bee colonies?

THE NATURE OF THE BEAST

As mites go, varroa is quite big, as most mites are invisible to the naked eye. The female varroa mites are the only ones normally seen. Each one is dark reddish brown with a smooth, elliptically-shaped 'shell' over its body and is 1.5–1.99 mm wide x 1.5–1.77 mm long. The mites are flattened front-to-back (dorso-ventrally), a shape which allows them to get between the overlapping plates on the bee's body, so making them very difficult to see. When they do crawl about, they remind me of little crabs. Their four pairs of legs are more or less hidden away under the 'shell'. They have claws on their feet and setae (hairs) on their bodies, which enable them

Female varroa mites
(top: underside; bottom: upperside)

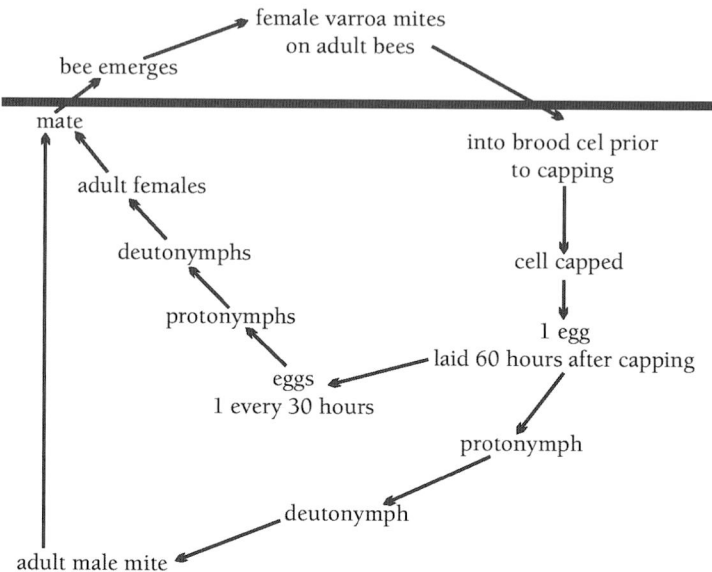

Life cycle of *Varroa destructor*

to cling onto the bee. Incidentally, because we usually see them dead or moribund (we hope) it is easy to forget that they are lively little creatures and, as well as walking, they can move very rapidly and can also jump. At the front end are the mouthparts which are adapted to pierce the outer covering of a honey bee prepupa, pupa or adult and suck the haemolymph (blood).

The life cycle

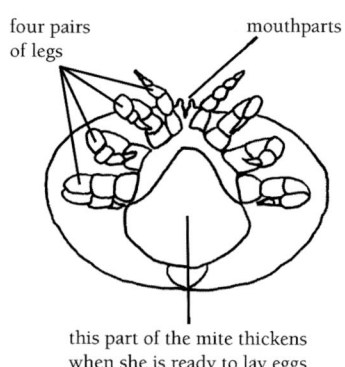

The female varroa mite (ventral surface)

In the active season each female mite hangs about on adult bees in the hive, probably for between 5 and 11 days. During this time she survives by feeding on the adult bee's haemolymph by piercing the membrane between the segmental plates which cover the bee This is called the ***phoretic*** (migratory) stage because the mite is transported to wherever the bee goes, inside and outside the hive, and so can migrate to other colonies. Movement to other colonies may be by drifting, robbing, invasion by colonies that are collapsing due to varroa, movement by beekeepers and the movement of swarms. It has been estimated that, when a swarm leaves its colony, approximately 25% of the mites leave with it on the bees, the remaining 75% staying in the parent colony, either on the bees or in the brood. When the colony is broodless, all the mites are phoretic and particularly vulnerable to treatments.

MIGHTY MITES

When there is brood in the hive, the mite is able to detect a brood cell that is about to be sealed. The method she uses is scent as there is a change in the substances given off from the surface of a larva as it matures so that, when it gets to a certain point in its development, the worker bees can smell the change and begin to cap the cell. At this point, our female mite uses the chemical message supplied by the bees for their own benefit, dives into the cell and submerges herself in the brood food at the bottom, extending two breathing tubes, called *peritremes*, up to the surface to enable her to breathe. Once the cell is sealed, about four hours after she entered it, she emerges from her hiding place and starts to feed on the haemolymph of the prepupa. The intake of haemolymph is probably essential to her ability to reproduce because, during this time, she increases in weight considerably and becomes much thicker. Then, about 60 hours later, she lays her first egg, on the side of the cell. This is unfertilised and always develops into a male. After that, she lays fertilised eggs at the rate of one every 30 hours. These all develop into females. Sometimes this system goes wrong and, like our honey bees, if the female mite fails to mate, or her mate is infertile, she lays only unfertilised (male) eggs. Each varroa mite lays approximately five eggs in total while in a brood cell.

An egg develops very rapidly before it hatches, after about half a day, into an immature mite called a *protonymph*. This grows and moults into a *deutonymph* which, again, grows and moults into

A male varroa mite on the head of the developing bee

Female varroa mites prefer drone cells as they can produce more daughter mites

the adult mite. Most other mites have a three-legged larval stage between egg and protonymph and a third nymphal stage after deutonymph but, because varroa mites have a relatively short time to develop, the life cycle is somewhat truncated: the brief larval stage occurs within the egg case and there is no third nymphal stage. Mites complete their development in between 5.5 and 7 days (figures vary depending on the source). The eggs and nymphs are all cream/white in colour until they darken with the final moult. All the nymphal stages, and mother, feed on the pupa's haemolymph but the young nymphs are unable to pierce the skin on their own so mother makes a hole and all the nymphs feed from this one point. The males remain in the cell, mate with all their mature sisters and then they die. (It seems to be even less fun being a male varroa mite than being a drone honey bee.) Where infestations are heavy and several mites enter a cell together, there will be some possibility of males mating with mites from another mother. It is possible, using a dissecting microscope or a hand lens, to see the males if you inspect infested drone brood which has been cut out. Then you will see that they are small and round in shape with very little colour.

The female mites, newly mated ones and mother, leave the cell with the hatching bee and the whole process starts again. A female mite may go through as many as four breeding cycles during its life.

Not totally successful

It is a strange fact of varroa life that many young mites die before they leave the cell. Why this should happen is a mystery. So, of the five eggs that a female varroa mite may lay theoretically in a worker cell, the fourth and fifth just do not have time to fully mature, but the third and even the second mites often die before leaving the cell, leaving an average of about 1.45 female offspring per cell. Similar figures apply to drone cells although the surviving numbers produced will be higher, around three, because of the longer post-capping time available for mite reproduction. (Worker brood is capped for 12 days and drone brood is capped for 14 days.) It is easy to see why mites 'prefer' drone brood. Of course, although the mite population grows slowly at first, it soon accelerates and then problems arise.

BUILD-UP IN THE COLONY

We have seen that the reproductive rate is quite low and, when varroa first arrived, some people were surprised that it could cause such devastation, but the build up is insidious. It is perhaps worth spending a little time just considering how a mite population grows. If we take 100 mites as a starting point, assume that each female mite in a worker cell produces, on average, 1.4 daughter mites and, in drone brood, 3 daughter mites and that each female mite goes through two breeding cycles before she dies (she may in fact go through as many as four cycles), then the numbers look like this:

Worker brood

100 ('starter' mites)
↓
140 + (100 mother mites) = 240
↓
336 + (140) = 476*
↓
666 + (336) = **1002**

This ten-fold increase would take about 50 days (less than two months) allowing 12 days for each breeding cycle and seven days between them. You can carry on with the arithmetic for yourself.

Drone brood

100 ('starter' mites)
↓
300 + (100 mother mites) = 400
↓
1200 + (300) = 1400*
↓
4500 +1200 = **5700**

* the original 100 mites have died

This time we are looking at a period of about 56 days because the drone pupa remains in the cell for longer, but this result is quite frightening. I have allowed for two breeding cycles per varroa mite but this may well be three or four. Of course, in a normal colony during the active season some mites will be reared in worker cells and some in drone cells, but there is some evidence to suggest that female mites will wait until the first drone brood is available in the spring, so giving the population growth an initial spurt. However we look at it, and taking into account all the imperfections of this simple scheme, 100 mites in early May can be very bad news indeed and can mean several thousand mites by the end of July.

EFFECTS OF THE MITE

The effect that the mite has on the individual bee, and subsequently on the colony, cannot be separated from the effects of the bee viruses usually present. Most beekeepers had never heard of bee viruses, with the exception of sacbrood, until the arrival of varroa: the whole subject appeared to be the exclusive province of research workers. Then, suddenly, they became very important.

The reason for this is that several viruses can be present in the bee without any observable effects but, once varroa has built up in a colony, this happy state of affairs changes. The mites probably activate at least some of the viruses already present and, because of their method of feeding, act as vectors, transferring viruses from one individual to another and, even more importantly, between adults and brood. It is now known that some viruses can also multiply in the varroa mites.

We discussed the individual viruses, and their effects on adult bees, in Chapter 5 so I am not going into them here, but the combined effects of varroa mites and the viruses in the developing pupa may result in:

- reduction in weight of the individual bee at emergence. This is due to the uptake of protein by the mite, so reducing the amount available to the bee for growth
- a shorter life-span for the individual bee
- lowering of the bee's immune response, so enabling other infections to take hold. This is a direct effect of the mite and sometimes, some of the viruses.

Where more than one mite is present in a cell, the weight loss will increase proportionately. This is clearly going to have an impact on the colony in terms of poorly developed and lost workers. Workers infested with more than one mother mite usually do not survive the pupal stage. In the case of drones, they may not survive the two weeks it takes them to reach sexual maturity after they emerge from their cells, or have such poorly developed muscles that they may never be able to mate. Stunted, completely useless individuals can be produced where there are many mites in an individual cell or the pupa may die.

On a colony scale the results can be catastrophic:

- in spring, the early deaths of many older bees, before sufficient young bees have been raised to take over from them, can lead to colony collapse
- where varroa is not well controlled, especially after resistance to a chemical control method has arisen, bees and brood may show a variety of symptoms, mostly associated with the viruses
- there may be complete colony breakdown. This may occur where a colony has been showing signs of disease, or without any previous warning. The hive may be left completely empty as the bees leave home, often invading other colonies nearby and taking their mites with them or, in some instances, crawling from the hive
- there may be a higher incidence of other, apparently unrelated, bee diseases due to the weakened state of the colony, the stress that varroa places upon it and the suppression of the immune response in the individual bee. (The term 'Parasitic Mite Syndrome' has sometimes been applied, usually in USA, to the complex of symptoms that can develop.)

CHEMICALS AND VARROA

Throughout the world, wherever varroa appeared, chemical treatments were quickly developed and used. These were usually very efficient at controlling the mite and beekeepers adapted to their use and breathed a huge sigh of relief. The table overleaf shows the treatments available which have been registered for use in the UK.

Other treatments, based on different chemicals, are registered for use in some other countries but not in the UK. Sometimes they are used in the UK by the National Bee Unit, under licence. In some cases they are very unpleasant chemicals.

The essential oils and formic acid all work (at least partly) by evaporation, so their success is dependent on temperature. If it is too cold they will not evaporate rapidly enough to be effective and if it is too hot they may evaporate too quickly and harm the bees, as well as not achieving the mite kill anticipated. Ventilation is also important, particularly in the case of formic acid so, when using MAQS® (Mite Away Quick Strips) it is advisable to leave the mesh

Period of application	Type of chemical	Trade name	Method of application
Spring or autumn	Synthetic pyrethroid	Bayvarol®	Hanging plastic strips
Spring or autumn	Synthetic pyrethroid	Apistan®	Hanging plastic strips
Spring or autumn	Essential oil	Apiguard®	Gel in packs or bulk
Spring or autumn	Essential oils	ApiLife VAR®	Tablet which has to be broken up
Late summer	Essential oils	Thymovar®	Tablets placed on the top bars
Spring/summer/autumn	Formic acid	MAQS®	Impregnated strips

floors open or, when solid floors are in use, give extra ventilation by offsetting the boxes a little. Care must also be taken when using thymol. In this case, the mesh floors should be closed by inserting the drawers beneath them, but the ventilation in the crownboard should remain open.

They are all chemicals and some are dangerous, both to bees and the beekeeper, so they need handling according to the manufacturer's instructions and always with care. All chemical treatments cause some degree of stress to the bees so they should be used only when necessary.

There is now evidence that many of the chemicals used in hives can build up in the wax. They are usually insoluble in water but dissolve in, or are absorbed by, oils and waxes. This can result in low levels of chemicals being present, even when the hive is not being treated for varroa, and can also produce a synergistic effect where different chemicals are being used at different times but, because of the presence of one in the comb, the effect of the second one can be more damaging to the bees: further good reasons for renewing brood comb frequently and destroying the old comb.

Resistance to chemicals

Many of the proprietary products, such as Apistan® and Bayvarol®, are very effective and very easy to use. Their development lulled many beekeepers into a false sense of security since they could treat their colonies once, or twice, a year with a product which was easy to use and apparently foolproof. However, after their use for a number of years, some mites develop resistance to the chemical. By resistance we mean that the target organism (the mite) is no longer susceptible to the normal levels of the chemical. Once resistance to a particular chemical has arisen among a mite population, the resistant mites begin to multiply within the colony, the chemical treatment no longer works and the beekeeper is back to the same situation as prevailed when varroa first became established.

Resistance to the synthetic pyrethroids appeared about 10 years after the mite had first been detected and resistance to coumaphos, amitraz and other chemicals licensed for use in some other countries, but not here, has now appeared in several countries.

How does resistance happen? I do not want to go into lots of involved chemistry here but, put simply, Apistan®, Bayvarol® and other 'hard' chemicals act in a particular way using, usually, one chemical pathway in the mite. In varroa mites resistant to the synthetic pyrethroid treatments, resistance appears to be due to two different mechanisms, not acting together but singly, and this suggests that resistance has developed twice at least.

- some mites may develop high levels of particular enzymes which break down the chemical before it gets to the site of action (this is called detoxification).
- there is always a pathway which enables a chemical to get to its site of action in the mite (or any other pest). If this pathway is changed, the chemical action is blocked. This type of resistance has occurred in varroa mites.

Once resistant mites are present they will become the dominant type of mite and will spread throughout the population. Under natural conditions this would be slow, but resistant varroa mites have wonderful friends in beekeepers who move colonies about, sometimes huge distances, may ignore the rules and frequently have no idea what is going on in their colonies.

Resistance to chemicals is not restricted to varroa mites but arises in many, many organisms, including bacteria, insects, lots of other mites and even mammals. Often the acquisition of resistance is at the expense of some other attribute. So, such organisms may not grow as fast or may not reproduce so well. Unfortunately, it appears that pyrethroid-resistant varroa mites do not suffer in this way and are just as vigorous as their pyrethroid-susceptible sisters. Thus they will still build up at the same rate and have the same effects on the colonies they inhabit but they will have one big advantage – they will survive when the familiar treatment is applied. That is quite an advantage (for them)!

A test kit for detecting pyrethroid-resistant varroa mites

Detecting resistance

There are two methods of discovering if an individual hive contains mites resistant to a particular chemical.

- treat as usual with the chemical, then count the number of mites falling after the treatment has been withdrawn. An effective chemical treatment should leave very few, if any, mites.
- carry out a resistance test. In the case of the synthetic pyrethroids this involves putting about 200 bees into a jar containing a small strip of Apistan®. The jar is covered with a plastic mesh, allowing the mites to fall through onto a piece of sticky paper underneath it. The jar is left in a dark place at room temperature for one to two hours. The number of mites on the sticky base is then counted, the bees are quickly killed in soapy water then rinsed through coarse and fine filters with lots of clean water and the mites which are washed off are counted. Those washed off are mites which have not been killed by the chemical and are therefore resistant. This gives a percentage of resistant mites but is only an approximation and does not work if there are very few mites to start with.

After resistance is established

The use of effective chemicals in varroa control often leads to complacency, so what happens once the chemical no longer kills the mites? There are three rules here.

- do not attempt to use the treatment to which the mites are resistant (although after a period of time it may again become usable).
- know how many mites you have in your hive(s). This requires monitoring, normally by having a mesh floor beneath the hive and collecting the mites which die naturally. This should be done at intervals throughout the year.
- do as much as you can to keep the mite population low. The target should be less than 1000 mites although in some other countries this figure may vary. This will involve various techniques which can be used at different times of the year, as well as the use of chemicals to which the mites are not resistant, when necessary. This approach is called *Integrated*

Pest Management (IPM) and is not a new concept, having been used frequently, particularly in organic gardening and farming systems.

The various techniques employed to reduce mite populations rely on three features of mite life:

- mites in their phoretic stage are exposed and therefore vulnerable
- mites prefer drone brood
- where there is a limited amount of brood containing fully grown larvae, mites are forced to use it. Such brood, once sealed, can then be removed.

NON-CHEMICAL CONTROL METHODS

1. *Use of mesh floors*. It is estimated that up to a 14% reduction in the varroa population can be achieved by the use of mesh floors. There must be a gap beneath the floor of at least 5 cm as, otherwise, mites will be able to detect the brood and crawl back up into the colony.
2. *Culling drone brood*. A brood frame, modified by horizontal wooden strips across the middle and filled with worker foundation at the top, or simply a super frame of worker foundation, can be inserted at the edge of the brood nest. The bees will usually fill the lower part with drone comb although I have seen several instances where they have built worker comb. The lower part of the frame can be fitted with drone foundation and some beekeepers use a whole frame of drone foundation. Once this brood is sealed, it can be removed from the hive and the frame returned to the bees who will repeat the exercise.
3. *Queen trapping*. This can rid a colony of virtually every mite if done properly. The queen is trapped in a cage covering a frame of clean comb (Frame 1) for 9 days. At the end of 9 days she is placed on a second frame of comb (Frame 2) in the cage and Frame 1 is marked and left in the hive. After a further 9 days, the queen is put on a third frame of comb in the cage, the brood on Frame 1, which will now be capped, is cut out and destroyed and Frame 2 is marked and put back in the hive. After another 9 days, Frame 3 is taken out of the cage, marked and put in the hive and the queen released back into the colony. The comb in Frame 2 is destroyed. After a further

Placing a frame of drawn comb, carrying the queen, in the queen cage

Putting the trapped queen back into the hive

9 days, the comb from Frame 3 is cut out and destroyed. This is an extremely effective method but relies on having drawn comb available and requires a reasonable amount of skill on the part of the beekeeper. It is also only really suitable for a small number of hives due to the labour involved. Its timing is crucial. It is vital not to start it before all the bees needed for the main honey flow are at least at the sealed brood stage and to complete it early enough to enable as many bees as possible to be reared for the winter period. These bees should be almost totally varroa free.

4 *Sugar dusting*. The jury is still out on the success of this procedure, but it may remove a proportion of the phoretic mites. To have any chance of making much difference, it will need to be used every time the colony is inspected. A fine mesh screen is placed over the brood box (or, for one or two hives an old kitchen sieve can be used), about a handful of icing sugar is spread over the screen and any lodging on the tops of the frames is brushed in. The mesh floor must be open so that the mites can fall well clear of the hive. The best estimate of the effectiveness of this method is that it may remove a small percentage of the phoretic mites and, used regularly, may led to a reasonable reduction but it is definitely not a control method on its own.

5 *The artificial swarm*. Bees can be artificially swarmed, the broodless part treated chemically and, if necessary, the first brood discarded, once it is capped. In extreme cases, where mite levels have risen alarmingly, all the sealed brood can be removed for a time. This is usually carried out in the autumn, on a colony that would otherwise die, and it is essential to feed it well afterwards.

Whatever methods the individual beekeeper employs it is always necessary to monitor them. This can be done in a number of ways.

- *natural mite drop* by counting mite numbers dropping through a mesh floor over one week or so. Danger numbers vary during the year from 0.5 mite/day in winter to 6 mites/day in May and 33 mites/day in August.
- *uncapping drone cells*. One hundred drone cells are uncapped with an uncapping fork. They should be at the pink-eye stage of development. Where 5–10% of cells are infested, action must be taken.
- *sugar roll* where approximately 300 bees are taken from the brood nest, put in a jar with a mesh lid and rolled with about a handful of icing sugar for two minutes. The jar is

then set aside in a shady place for one to two minutes and the sugar, and mites, is then shaken out through the mesh into a shallow dish of water set over a white surface. The rolling and shaking is repeated twice more and the bees are then released in front of their hive. The number of mites shaken will only reflect the number on those 300 bees. So, if five are shaken off and the hive holds approximately 30,000 bees, that will give a total of about 500 mites. Added to this, whenever there is sealed brood in the hive, perhaps 50–70% of the mites will be in that brood. So, five mites will give a grand total exceeding the 1000 mite threshold and requires action.

Brood removed for varroa control must always be sealed and increasing the amount of drone brood in the colony can act as a wonderful boost for varroa if it is left to emerge. We all hope that, at some stage, biological treatments or varroa-resistant bees will become available.

LONG-TERM SOLUTIONS?

Attempts are continuing in several parts of the world to develop a bee that is resistant to varroa. Of particular interest are hygienic bees. These are bees that can recognise when brood is diseased in some way, uncap it and remove it. It used to be thought that this was a simple genetic trait, but it is now obvious that it is not. It is dependent on the ability of the worker bees to detect the damaged larvae/pupae, it is thought by smell, so there are far more genes involved than was originally believed. Breeding of VSH (Varroa Sensitive Hygienic) bees is also being undertaken in several countries including the UK. These bees will selectively remove pupae infested with varroa.

Although the idea of a resistant bee is very seductive, breeding programmes aimed at one particular characteristic must be undertaken with care, as other traits can be lost in the process. So production, docility and resistance to other diseases may be affected. There is also the problem of persuading beekeepers, notoriously individualistic people, to use a particular type of bee.

ANOTHER POTENTIAL PROBLEM

Varroa is not the only mite which parasitises honey bees. It has a cousin with the delightful name of *Tropilaelaps*, at the moment absent from the UK, but a potential threat. There are several species but they can be simply referred to by the general name. It is much smaller than varroa and its natural host is *A. dorsata*, but it can parasitise other *Apis* species including *A. mellifera*. It is a creature of the tropics and sub-tropics and it is thought that it cannot survive a break in brood rearing in a colony, so it may not be able to survive in temperate climates where the queens stop laying for a time. It is a notifiable pest in the UK.

7 THE ENEMY WITHIN AND WITHOUT

A hive of honey bees is an attractive potential home or food source for many other creatures, some of them smaller than the bees that inhabit it, but many of them bigger, so we are going to look at this list of potential problems in order of size.

ONE OF LIFE'S LITTLE PUZZLES!

At just over 1 mm wide and less than 2 mm long from its nose to its tail, *Braula coeca* is a tiny enigma. It is a fly which cannot (fly that is), is often mistakenly called a louse (not in the insulting sense you understand), half of its name is pronounced completely differently from the way it is spelt and it is said to be a harmless inhabitant of beehives. But is it? It belongs to the enormous group of insects called the *Diptera* or two-winged flies, but it has no wings at all. Living exclusively within a colony of honey bees it does not need any wings because it does not fly anywhere. Its common name is the bee louse because it bears a close resemblance to the insects in the groups Mallophaga and Anoplura. These are biting and sucking lice respectively and are parasites on birds and mammals. To the beekeepers who gave *B. coeca* its common name, these insects would have been very familiar and they assumed that the similar looking insect, which crawled over their bees, was a louse.

Life cycle

In its early stages *B. coeca* shows all the characteristics of its normal fly relatives. It starts life as a tiny egg which hatches into a legless maggot about 1.5 mm long when fully grown. It eventually pupates inside its last larval skin which is then called a puparium. This is very characteristic of large groups of the Diptera. The egg is laid

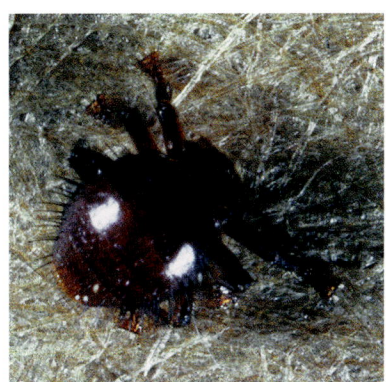

Braula coeca

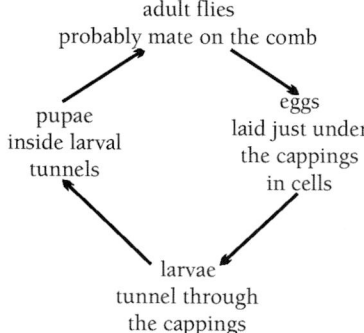

Life cycle of *Braula coeca*

by the female in the cells containing honey, just as they are being capped and, after hatching, the larvae tunnel through the cappings, feeding on honey and pollen. They pupate inside the tunnel and, when the adult fly emerges 21 days after the egg was laid, it climbs onto the body of a bee. This is where the special adaptations of this unusual insect become important:

- its body is covered with stiff hairs, helping it to hold on
- its feet are equipped with stiff spines to enable it to grip the hairs of the bee
- its mouthparts are adapted so that it can obtain food from the mouthparts of the bee when the bee is feeding or exchanging food with other bees. It perches at the front of the bee's head and reaches into the cavity at the base of the glossa.

The bee louse, the bee and the beekeeper

B. coeca is described as an *inquiline* in bee nests. An inquiline is an organism which lives with another species without harm to either. As far as it goes that is fine. *B. coeca* is not a parasite, does not bite bees, suck their blood or spread their viruses. But it seems to me that it must act as an irritant and a nuisance, if nothing more, and this applies particularly to the queen, because she is very attractive to *B. coeca*. I have seen queens with their heads and thoraces covered by a mass of crawling *B. coeca* and no one can convince me that such a queen is not at least irritated by their presence and is therefore likely to be less efficient. To the *B. coeca*, the queen is attractive as the most stable permanent member of the hive community and because she is constantly being fed by other bees.

For the beekeeper, the main problem is those tunnels. Because they are confined to the cappings, extracted honey is fine, but if you are producing cut-comb the little wiggly white lines are not good news. Inside them are larvae and pupae and who wants to eat honey with maggots in it? The usual advice is to put the comb in the freezer for 24 hours, but then it will merely be comb with dead maggots in it!

The good news

Since the advent of varroa and the use of acaricides, the numbers of *B. coeca* in hives has plummeted. I used to have lots and now I rarely see them. Personally, though, I would rather have put up with the minor problems caused by *B. coeca*.

THE ENEMY WITHIN AND WITHOUT

SMALL HIVE BEETLES

At the time of writing, Small Hive Beetle (*Aethina tumida*) has not been recorded in the UK, but rest assured that it will probably come. Small hive beetle (SHB) will make mice and wasps look positively benign.

It is small, only about 5.7 mm in length, black and with tiny clubbed antennae. Each female beetle can lay up to 1000 eggs, hidden away in crevices in the hive or laid in comb containing pollen or brood. These hatch, after a few days, into tiny larvae which feed on bee eggs and larvae, pollen and honey, tunnelling through the wax in the process. They can live in stored boxes as well as active hives. Their faeces get into the honey, causing it to ferment and become frothy and unusable, even for bee feed. There is no webbing, as with the wax moth larvae (see below), but the combs become slimy and the beetles, in addition to weakening a colony and depleting it of brood and/or stores, can become a serious threat to survival.

Once fully grown at about 10 mm and after two weeks or thereabouts, the larvae start wandering about and leave the hive, finding a suitable site, in the soil, to pupate. The adult beetles then emerge after three-to-four weeks and enter hives. Development times depend very much on the weather and, if the soil temperature is below 10 °C for an extended period, the pupae may not be able to complete their development, so our climate may be our saviour in this case.

Adult small hive beetle (*Aethina tumida*)
Note the clubbed antennae

Small hive beetle larvae
Note the three pair of forelegs

A larva pupating in sand

Stages of small hive beetle pupation

Moving about

SHBs can fly quite long distances so, once in an area, they have no trouble moving about. As well as entering a country in beehives and on bees, the pupae can also be transported on soil round the roots of plants. The adults can eat fruit, particularly melons, so this forms another means of entry into a new area. The species originated in sub-Saharan Africa but is now causing big problems in the USA so its ability to spread great distances should not be underestimated. It is a notifiable pest in the UK.

WAX MOTHS

Most beekeepers are only too familiar with wax moths, but there is confusion concerning their life cycles and even their identification, often not helped by reading the various accounts in bee books.

Galleria mellonella
The Greater Wax Moth

Achroia grisella
The Lesser Wax Moth

In some of the older books they are referred to as 'wax worms' which is a complete misnomer. So what are wax moths, how do they live, why do they cause problems for the beekeeper and what can be done to control them?

What are wax moths?

The term 'wax moths' normally refers to those species of moth whose larvae live in the comb made by honey bees, causing damage in the process. (There is a separate species which affects bumblebee nests and the honey bee wax moths may also affect bumblebees and nests of stingless bees.) They are found wherever the honey bee exists and are more troublesome in warmer climates. In the great scheme of things, wax moths have an important function in destroying the combs of feral and neglected colonies which have died from American foul brood (AFB) and other diseases, thus removing the disease organisms.

In the UK there are two species, the Greater Wax Moth, *Galleria mellonella* and the Lesser Wax Moth, *Achroia grisella*. They both belong to a section of the Lepidoptera (butterflies and moths) called the Microlepidoptera, an artificial group containing hundreds of small species of moths, many of them difficult to identify or tell apart. The group includes some serious economic pests such as the clothes moths and other species which attack stored food products, the remainder generally feeding on parts of plants.

The adult moths of the two wax moth species are quite different in appearance, when newly emerged: *G. mellonella* females measure about 28 mm (range of 25–35 mm) from wing tip to wing tip, although size is variable and relates to nutrition of the larvae. The moths have mottled buff and brown forewings with paler hind wings. A characteristic is the concave shape of the outer edge of the forewings, particularly obvious in the male. *A. grisella* is smaller, measuring approximately 18 mm (range of 15–20 mm) across its outstretched wings, which overlap slightly when it is at rest. It is a silvery grey colour, but the front of the thorax and the head are paler and appear almost white. Older moths of both species usually do not retain their colouring as the scales coating the wings are worn off. Both species may then just look like drab, buff-coloured moths.

Life cycles and behaviour

The life cycles of the two species are similar: Soon after hatching,

THE ENEMY WITHIN AND WITHOUT

the female moths leave the hive. The males produce a pheromone and an ultrasonic sound to attract the females. (This is very unusual in the Lepidoptera as the female usually attracts the male.) After mating, the females enter a hive or box of unoccupied comb in which to lay their eggs. They are able to enter through very small cracks and usually gain entrance to active hives at night.

Each moth lays a large number of creamy white eggs, several hundred usually, depositing them in batches in cracks and crevices in boxes and combs. After a few days they hatch into larvae, which are a dull white colour with a reddish brown head. The time spent as a larva varies between the two species. In *G. mellonella* it is about four weeks whereas in *A. grisella* it is between two and three weeks. The rate of development is very dependent on temperature. Inside an occupied hive, the high temperature will enable the moths to develop in the normal time but where they have to endure lower temperatures, the larval state may persist for several months. The larva is able to digest wax but its favoured food is the old larval and pupal skins of the bees and pollen. In fact, wax moth larvae living in very clean comb, which has never been used by the bees for breeding and which contains no pollen, may not be able to complete their development but remain perpetually in the larval state.

The larvae of the two species are difficult to tell apart but one distinguishing feature is their behaviour when disturbed. *G. mellonella* tends to run whereas *A. grisella* feigns death and remains still.

Once the larvae are fully grown they change into pupae, each enclosed inside a silken cocoon.

After about a week, the new generation of moths hatches and the cycle starts again. The whole life cycle at an optimum temperature of 35 °C may be as little as four weeks for *G. mellonella* and even less for *A. grisella*.

Wax moths and the beekeeper

The larval stage is the one which causes problems for the beekeeper as the larvae of both moths tunnel through the comb. As they do so, they surround themselves with a silken tunnel to which their faeces and bits of wax become attached, so making a thorough mess. In time, the whole comb disintegrates into a mass of silken webs and frass (faeces) and is completely unusable, involving an economic loss to any beekeeper.

Further problems occur when the larvae of *G. mellonella* prepare to pupate. They excavate hollows in the woodwork, and may even

A wax moth larva

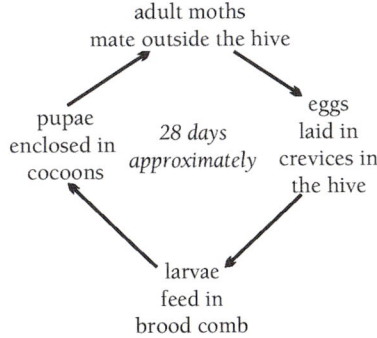

Life cycle of *Galleria mellonella*
(That of Achroia grisella *is shorter)*

The larvae of lesser wax moths (*Achroia grisella*) gather together to pupate inside cocoons
The holes show were the adult moths have emerged

make holes in frames, where they spin their cocoons. To add to the problem the larvae, at this stage, become gregarious and tend to line up in rows, very close together, so that considerable damage can be done to frames and hives. *A. grisella* does not have this annoying habit and usually pupates on the comb, its cocoons covered in frass so that they appear dark.

Most problems occur in unoccupied boxes of comb, especially those stored away during the winter period but, increasingly, damage is seen in colonies during the active season. Colonies which have been weakened by varroa or other diseases are particularly at risk and, in the worst case, the combs can be reduced to a mass of webs and cocoons, all fastened firmly together and completely unusable.

Additional problems

Sometimes a single larva, tunnelling through sealed brood, will result in the condition known as bald brood. The bees remove the cappings, damaged by the larva, to expose the bee pupae beneath. This, in itself, does not seem to cause any harm, but the presence of the faeces of the moth larva results in deformations of the legs and wings in the developing bees.

Comb honey is occasionally affected, usually by the young larvae of *G. mellonella*, the damage appearing as silvery tunnels in the cappings. It is recommended that sections and portions of cut comb are put in the freezer for a few days to destroy the larvae.

Control

In active colonies control is usually achieved by maintaining strong colonies with a minimum of unoccupied comb. Where colonies are weak, a dummy board should be used and superfluous frames removed. Any larvae or moths discovered in a hive must be destroyed although the adults are notoriously difficult to catch as they dart about.

In stored comb a number of options are available. Boxes which are stored outside and subjected to periods of freezing temperatures will be safe although, with warmer winters, this may not be reliable. The old books list various fumigants to treat combs. These are now regarded as unsafe and are mostly unobtainable.

Ethanoic (acetic) acid

This will kill most stages of wax moth, except those larvae which are deep within the comb. It is used, at a concentration of 80%, as a

THE ENEMY WITHIN AND WITHOUT

fumigant in stacks of boxes. It should be used outside. Metal parts, such as metal spacers and runners, must be removed or coated with petroleum jelly. The acid must not come into contact with concrete and should be handled with great care (see page 35). Rubber gloves and goggles must be worn when handling the acid and the boxes must be well aired before use.

Bacillus thuringiensis

This bacterium is a form of biological control. It is a specific parasite of lepidopterous larvae and can be sprayed onto the combs as a suspension. However, it will only kill young larvae so is used as a preventative treatment for stored comb. It is harmless to the beekeeper and the bees and its effects are said to last into the active season. It can be obtained from the bee supplies companies and its only disadvantages appear to be its cost and the time taken to spray the combs. A word of warning here, from personal experience: the frames must be dried off very thoroughly before being stored away as, otherwise, moulds may form on them.

Temperature

Subjecting combs to low temperatures for a period will kill the eggs and larvae. Minus 15 °C for two hours is adequate but longer will be needed for thick combs or whole boxes of combs. A domestic freezer is suitable, but its use may be a matter for discussion with others in the household – sometimes lively!

The future

Since the removal from use of chemical methods of control for wax moths, many beekeepers have experimented with various herbs, including lavender, with some success. As in many aspects of beekeeping, ideas and methods are changing and it is best to keep an open mind and, above all, be prepared to adapt. If all else fails, stop keeping bees and breed wax moths – there is a demand for them as bait amongst the angling fraternity.

WASPS

These are truly wonderful creatures. As we have seen in Chapter 1, they are closely related to our bees but their larvae are carnivorous. For that reason they are friends of the gardener for most of the year, consuming a large number of little larvae and other insects, many

Wasps may target a colony, particularly a small one, in late summer and rob it of its stores

of them pests, but, once the adults no longer have any larvae of their own to feed and their supply of sweet larval saliva finishes, they go in search of other sweet foods and a weak colony of honey bees can become a target.

The inhabitants of a wasp nest will often concentrate on one hive and, by repeated attacks and sheer weight of numbers, will gain access to the honey stores. A colony can lose all its winter stores in this way and many bees may die attempting to defend the colony. Sometimes it is possible to open a hive in this condition and find enormous numbers of wasps inside, plundering the stores. The bees so attacked tend to give up the fight and small colonies such as nuclei are particularly at risk.

The wasps behaving in this way are usually the two species of small common wasps, *Vespula vulgaris* and *V. germanica*, but occasionally there are reports of hornets, *Vespa crabro*, attacking honey bees. This does happen sometimes, as the hornets may find the bees easy prey and use them to feed their larvae, but the problem is not major and rarely results in much damage. It is a different story in some tropical and sub-tropical parts of the world where the hornets are much bigger and can be responsible for the deaths of large numbers of honey bees at the nest entrance, often completing the destruction by moving into the hive and killing the remaining occupants and the larvae. This is a particular problem in Japan where many colonies are lost in this way.

What can we do about wasps?

Reducing hive entrances to a minimum, one bee space if necessary, enables the bees to defend their home. This is the first line of defence and is best done late in the summer before wasps become a problem. Some beekeepers destroy as many wasp nests as they can and kill queen wasps in the spring, many of which may hibernate under hive roofs in the winter, but this seems rather an extreme measure when wasps are only a problem for a short time in late summer and autumn. Better to leave them until later in the year when the queens and males have left. Nests destroyed at this stage will have fulfilled their purpose. Some beekeepers use wasp traps, siting jam jars half full of beer near the hive entrances or purchasing commercially produced traps.

THE ASIAN HORNET

The Asian Hornet (*Vespa velutina*) is just another wasp, but as it is another problem-in-waiting at the moment I have given it a separate section. It was introduced into France at some point and

Vespa velutina

THE ENEMY WITHIN AND WITHOUT

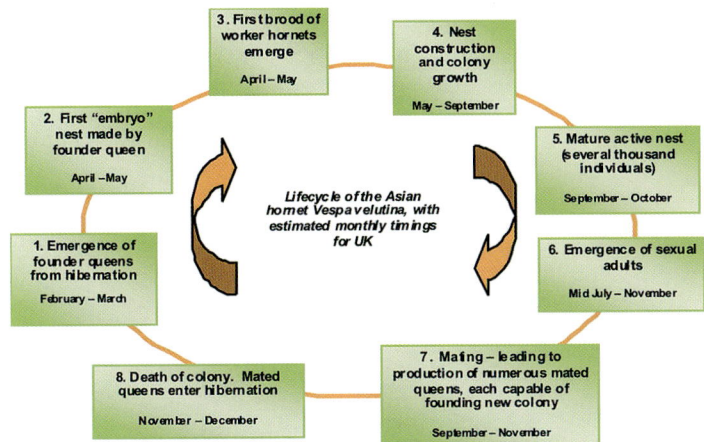

Vespa velutina life cycle

has spread rapidly, invading some neighbouring countries. Poised just across the English Channel, it will arrive at some time, either under its own steam or hidden in a consignment of some sort. It is smaller than our native hornet and is almost entirely black with one yellow/orange band round its abdomen. Another diagnostic feature is the yellow tips to its legs. It forms huge nests, usually quite high up in trees, and can target hives, killing the bees at the entrance, weakening the colony and, sometimes, causing its demise. In the UK, suspected sightings should be reported to GB Non Native Species Secretariat (www.nonnativespecies.org or nnss@fera.gsi.gov.uk) and it is also helpful if the National Bee Unit (nbu@fera.gsi.gov.uk) is informed.

MICE

Mouse damage is usually due to the Wood Mouse, *Apodemus sylaticus*, but may also be caused by the House Mouse, *Mus domesticus*. It is true to say that it is always avoidable but then we do not live in an ideal world. Mice usually gain access to active hives in the cooler days of autumn when they are small and the bees are forming a cluster. Once inside the hive, they feed on the stores of honey and pollen plus adult bees and any larvae, disturb the winter cluster and usually make a nest, damaging comb and frames in the process. Pieces of comb may appear outside the hive entrance. Honey bees dislike the smell of mouse urine and droppings and will ignore the part of the hive occupied by the mouse.

Mice can be excluded from occupied hives by ensuring that entrances are too small for them to get in, either by using a narrow

A narrow entrance, 8 mm high or less, will keep out mice

CHAPTER 7

A mouse nest in a disused super
Mice can do greater damage inside occupied hives during the winter

slot, 8 mm high, or fixing a mouseguard over the entrance. These can be purchased or made but it is important to realise that bees passing through the holes can lose many loads of pollen, so mouseguards must not be put on too early in the autumn and must be removed before the bees start bringing in pollen in the spring.

Mice can also be a problem in stored equipment but again they should simply be denied access. Equipment stored in a mouse-proof building should be safe, but supers and other boxes stored outside or under shelters provide ideal homes for mice. The empty brood boxes and supers must be sealed to prevent the mice finding a way in and mouse populations should be reduced by the use of traps or baited rodent traps, taking care that non-target species are not killed.

Mice need to be taken seriously although they can easily be controlled, but always remember that they are very small and can squeeze through tiny holes.

WOODPECKERS

These beautiful birds, as their name suggests, peck wood and can be wreckers of beehives. The normal culprit is the Green Woodpecker,

THE ENEMY WITHIN AND WITHOUT

Picus viridis, which can attack beehives in the winter, in search of bees and larvae to eat. It has a very powerful beak which can bore holes in beehives very rapidly. (In fact, woodborer might be a better name for this creature.) Its tongue can reach several centimetres beyond the end of its beak, is sticky and barbed and the bird uses this to pull bees out of the hive.

Besides the disturbance and loss of bees, the main problem is the damage to the hive bodies, which can end up with big holes in them, rendering them no longer weatherproof and allowing in other animals such as rats and mice. Colonies may easily be lost due to the attentions of the woodpeckers.

The strange fact is that Green Woodpeckers and hives can exist happily side-by-side for years. This is the case in my own garden where I had hives and Green Woodpeckers for more than 20 years before I encountered any problems. The birds spend hours feeding on the ants and their larvae, which seem to form a continuous ant nest along our driveway, and even bring their babies to visit us. However, in one hard winter period, with the ground frozen and natural food scarce, one of the birds discovered the hives. Unfortunately, as woodpeckers seem to communicate with one another, a continuous onslaught is inevitable and steps have to be taken.

In the UK the birds are protected by law, preventing them from being killed or disturbed in any way, so barrier control is the key with the usual recommendation that hives are covered by wire netting, leaving a space between the hive and the netting and ensuring that the woodpeckers cannot get inside the netting.

The green woodpecker (*Picus viridis*) loves ants, but may decide bees and their grubs are more attractive during cold weather

Woodpecker damage to a brood box

AN ONGOING STORY

There are lots of other birds and animals which might take a few honey bees as food and here we might include, as examples, flycatchers, tits, hedgehogs and toads, but these are never more than very minor skirmishes and often the bee comes off best! Damage can be caused by badgers and rats. The former can wreck a colony, particularly a small one, and I have seen the damage that they can do, but I have lived with them, both at home and in out-apiaries, with no more than a few scratches on hive boxes. Just be grateful that, in the UK, at least we do not have bears roaming around reducing apiaries to matchwood overnight.

SUMMING UP

The last five chapters have been a rather depressing account of the problems that may beset a colony of honey bees and it is important to bring some of this together and look at the problems as a whole rather than as individual ones. I have attempted to do this in the next chapter. In the USA, there has been the problem of Colony Collapse Disorder (CCD), a condition which has had a great deal of research and money directed towards it, still with no definitive answers. It is likely that it is due to a number of different factors interacting together and, always, varroa is likely to be at the heart of the matter. Its ability to depress the immune system of the bee and to transmit virus diseases makes it a very dangerous creature.

Our bees cannot avoid the many perils both inside and outside the hive but, as beekeepers, we can be aware of the problems and try, at all times, to alleviate as many as possible, particularly by controlling varroa, keeping strong colonies with young, productive queens, providing them regularly with clean comb and hives, and ensuring that they are not stressed further by shortage of food and by our treatment of them.

8 THE BROADER PICTURE

Keeping bees used to be comparatively simple: get a colony of bees, control swarming (I did say comparatively simple), remove the honey when ripe and make sure the colonies have sufficient food to last the winter. Crops of honey were generally good and the bees, on the whole, remained healthy. In older bee books the chapter on disease was usually to be found at the end of the book and the tyro beekeeper often ran out of steam before (s)he got there. That did not matter greatly because (s)he rarely saw anything serious. Then, in 1992, varroa arrived and, in the next few years, the beekeeping landscape changed. In the intervening years, there have been huge developments and we have reached an interesting time for bees and beekeepers. In this chapter I have attempted to summarise the problems that bees face. These are not restricted to the UK and this chapter is somewhat dismal, but it does bring to an end the portion of the book on the ills that befall bees. Just bear with it and Chapter 9 will cheer you up.

THE CURRENT SITUATION

For the past few years, losses of colonies during the winter have increased. This has occurred in many countries, not just the UK, and has given rise to mass panic in some sections of the media, which predict the demise of the honey bee. Because of the huge publicity surrounding bees, many people have taken up beekeeping as a hobby, but this should not mask the fact that many older beekeepers have given up. It is also a sad fact that many of these new beekeepers, who begin with such enthusiasm and often see themselves as saviours of the environment, give up after a very short time, unable to cope with the many challenges now facing beekeepers. A further problem is also arising in some areas where there are too many hives for the available forage. We will come back to that.

So why are bees in trouble and should we be concerned? After all, honey bees have been around for a very long time and have been through difficult periods before.

The role of varroa and viruses

As we saw in Chapter 6, varroa is a very serious pest which has spread around most of the world and caused problems wherever it has appeared. A large number of beekeepers simply gave up shortly after its arrival and, typically, in most countries it resulted initially in a 40% drop in beekeeper numbers. New beekeepers, who had never known life pre-varroa, adapted to dealing with it and, of course, it is now part of routine management, although that is not to diminish its effects. The ability of varroa to spread bee viruses and to depress the immune system of the bee, enabling other diseases to move in, are the two main reasons for its importance. It, together with the associated viruses, is still causing the deaths of many colonies worldwide. New viruses have been discovered and no doubt others are waiting round the corner, and we must not forget that viruses continually change so that there is constant competition between the virus and its host.

Varroa has resulted in the use of more chemicals in hives and this can be a cause of bee losses, damage to brood and problems with queen and drone development and function. Many of the chemicals used are absorbed into the wax and there are synergistic effects between some of them.

There is no doubt that, although conditions and diseases in different parts of the world vary, varroa, and sometimes its treatment, forms a base upon which the rest of the problems develop and build.

Other bee diseases

The diseases which move in as a result of the bees' lowered immunity cause still further problems and here nosema must be mentioned, particularly *N. ceranae*, which is still, to some extent, an unknown enemy. It has probably been around for longer than anyone knows but has only been identified as a different species comparatively recently. Both species of nosema are gut parasites, although *N. ceranae* may move outside the gut, causing greater damage to the gut wall than *N. apis*, thus allowing entry, via this route, for other viruses and bacteria. The two nosemas also reduce the available nutrition for the bee and place on it a demand for more energy. All these effects are going to be deleterious.

Fortunately, the foul broods do not seem to be increasing to any degree in the UK although European foul brood (EFB) is a problem in some areas and, the more we learn about the causative organism, the more we realise how complex it is.

In some cases, colonies can become so diseased that it is difficult to decide exactly what is wrong with them and a term used to describe this condition, although not commonly in the UK, is Parasitic Mite Syndrome. It describes a complete breakdown in the colony with death of brood and bees. The fundamental cause is varroa with other diseases often present as a result.

Loss of habitat

This is a longstanding and ongoing problem and is causing immense problems to all species of bees (and to most insects and many other animals). The increasing human population and spread of urban areas reduces the space available for wild creatures. The urbanisation of many places, which were once semi-wild, and the general desire for tidiness, is removing much needed forage. On the brighter side, more people are now aware of the value of bees and are more protective of them and there is recognition of the value of gardens, with more interest in conservation gardening and the provision of nectar-providing flowers. There is still a long way to go here though and the great British public still seems, on the whole, to be supportive of bees in a general sense, while not being quite so keen when they are up close.

A further problem for both solitary bees and bumblebees is loss of nest sites, although this clearly does not affect honey bees.

In rural areas, forage can frequently be scarce for a number of reasons, many of them discussed below, and in some urban areas the increase in the number of new beekeepers has led to a greater concentration of hives, reducing the amount of forage available to each. This competition for space and food can also result in an increase in some diseases.

Changes in farming practices

Agriculture has changed out of all recognition in the past 50 years. It is easy to criticise farmers but we must remember that there is an ever-increasing population to feed, in the UK people seem to think that food should be cheap and we are becoming heavily dependent on imports of food, often from countries which could do with it for their own populations. Many farmers have gone out of business

A hedge-laying competition. Treating a hedge in this way preserves the species there as well as providing a stock-proof boundary

Many miles of hedges have been removed, reducing forage and nest sites for wild bees. This trend has now been reversed, but much harm has been done

Greater efficiency and use of bigger machines have, in places, created agricultural 'deserts' with no place for wild life or bees

and others are controlled to a large extent by the large supermarkets and their pricing policies.

We can list some of the major changes to agriculture that have taken place:

- much bigger farms to enable economic survival
- reduction in the number of people working on farms
- bigger fields to allow use of larger machines
- removal of hedges and/or different maintenance methods for them
- greater use of nitrogenous fertilizers
- more intensive units for livestock production
- growth of crops for non-food use, eg. as biofuel
- use of different crops
- use of agricultural chemicals.

The first four on the list are all interconnected. Larger farms, employing fewer people and using bigger machinery, does not seem a problem on the surface, but it underlies massive changes to bee habitat. Many miles of hedge have been removed, both to allow access for larger machines and to maximise the acreage available for crops. Hedges, and the area at their base, are important sources of food for bees, as well as providing nesting places for wild species and allowing different populations of wild bees to move about so that the gene pool can remain robust. Hedgerow plants such as blackberry, hawthorn, blackthorn, dog rose, guelder rose and many more provide nectar and pollen, while many of the plants growing at the bases of hedges are also important.

Another problem which has arisen from the reduction in the agricultural workforce involves the maintenance of hedges: hedge-laying and trimming and ditch clearance, used to provide winter employment for a large workforce of full-time agricultural workers, but now machinery is used to cut hedges, rather brutally quite often, and frequently just as flowering plants are in bloom. This is damaging in the short term and, in the long term, hedges treated in this way frequently start to become 'gappy' and reduced. They are then grubbed up to be replaced by fences.

Large quantities of nitrogenous fertilizers have been used to promote growth and increase yields and this has led to a reduction in the use of clovers as nitrogen fixers. This is probably changing slowly as the high cost of fertilizers is limiting their use and farmers are using less. However, in some other parts of the world, use of nitrogen on a massive scale is causing a great deal of damage to the environment in general and, in particular, to water sources.

We are all aware that flowering meadows have largely disappeared.

THE BROADER PICTURE

Small fields full of flowers and traditionally cut for hay, can still be seen in the Yorkshire Dales

This has been the case for more than 50 years and they have become a rare site, restricted to a few places such as the Yorkshire Dales. Animals graze mostly on temporary grass leys, which are more productive than permanent pasture and are ploughed and reseeded every three or four years. Much grass conservation is as silage, cut when the grass is at its nutritional peak, rather than as hay, which encourages the growth of wild flowers. Even more extreme is the development of huge intensive animal units which sometimes use no grazing at all. Some beef enterprises fed primarily on barley come to mind. The overall result is more land under arable cultivation (and the importation of more animal feed).

Although the land devoted to growth of cereals has remained fairly static over the past 50 years, other crops have changed. Some of these changes have been good news for bees and beekeepers. There is a great increase in plants grown for their oils and, from the beekeeper's point of view, the most important of these is oilseed rape. Although it causes extraction problems, a great deal of honey is produced from it. Linseed and sunflowers seemed potentially good but, in my experience, the former does not produce any honey, although the bees visit it during the early part of the day before the petals fall, and the latter has proved so unpredictable in our climate that it is restricted to southern and eastern counties where it is drier and the summers are longer.

CHAPTER 8

A comparatively new crop, borage can produce enormous crops of honey

The attractive flower of borage has given the name Starflower to the honey

Oilseed rape produces a very good supply of nectar and pollen, but for a comparatively short time

Borage is a different matter and is a bee plant *par excellence*. It produces enormous crops of honey, which granulates very slowly, but is quite insipid and contains so much sucrose that the Honey Regulations had to be changed slightly so that it would comply with them. Borage is grown under contract and the amount available in a particular area fluctuates in different years, but where it is grown, it is very well worth moving bees to it if the beekeeper wants to boost his honey crop. The flowers are also very pretty.

It would not be possible to leave this section without a mention of field beans. There is currently between 110,000 and 170,000 hectares of these being grown in the UK. The seed is high in protein and used mainly for animal feed, reducing to some extent the reliance on imported soya. Good crops of honey can be obtained from beans although the older beekeepers tell me that the crops were better from the varieties being grown 50 years ago. Rose-tinted spectacles maybe, but it is entirely possible.

Crops are also being grown to produce biofuel. Many of these are non-flowering grasses or woody plants but some oilseed rape is sown for this purpose.

No doubt other novel crops will be developed as time goes on, some of them useful for our bees, others not. Within one crop type, varieties can vary tremendously as to their usefulness to bees and we have to remember that nectar production is not something which occupies the minds of most plant breeders. They are more interested in growth rates, yield, nutritional value, pest and disease resistance, ease of harvesting and many other important attributes. In some situations, self-fertile varieties are sought so that bees are no longer essential for pollination. This is happening in the USA where a self-fertile variety of almond is being developed with the aim of reducing the bee-dependency of the very important almond industry in California, following the loss of so many colonies from so-called Colony Collapse Disorder (CCD). What must always be borne in mind is that some agricultural crops produce a great deal of honey but flower over a short period of time. Bee colonies need food throughout the active season, so it is vital to have other sources available.

AGRICULTURAL CHEMICALS

Finally, in this section, we come to the use of agricultural chemicals, particularly those used for the control of plant diseases, especially fungi, and insects. Here again we must bear in mind the need for farmers to produce good crops and to protect those crops from attack by diseases and pests. In an ideal world no chemicals would be used, but this is a counsel of perfection and we live in the real world where aphids, caterpillars, beetles and many more insects, not to mention the rusts, blights and so on of the fungal world, can devastate crops.

Fungicides

If we can look briefly at fungicides first: these were never considered as a problem for bees until recently when it has been demonstrated that they are not as blameless as previously believed. They do not kill adult bees, as far as we know, but the effects on the colony are insidious. Some fungicides get into the bee bread, which bees produce from the pollen they collect. This can seriously harm the fungi which are important constituents and may result in bee bread which is of poor nutritional value or, in some cases, directly harmful. Poor bee bread, eaten by the nurse bees to manufacture brood food, will lead to poor nutrition for the larvae. The effect may be delayed and can be persistent as poorly nourished larvae become poor nurses for the next generation.

Some fungicides have also been found to be synergistic with some acaricides used in hives to combat varroa. The effects of the varroa treatment on the bees will be magnified and more deaths may result. We always have to remember that, although acaricides are aimed at mites, insects and mites are very similar and anything which upsets the dose is liable to cause harm to the bees.

This illustrates a general point: chemicals do not operate in isolation but are part of a much bigger picture, often working together or producing unforeseen results when found in association with one another.

Insecticides

That brings us, finally, to insecticides. These have been developed over a long period of time and are all subject to rigorous testing before being released for use, but the fact remains that they are all designed to kill insects – and bees are insects. There lies the dilemma. Insecticides such as DDT, developed during the 1940s, revolutionised agriculture but caused enormous harm to the environment in the process. Different generations of chemicals were developed with differing effects on beneficial insects. Stringent Codes of Practice were introduced, aimed at ensuring that chemicals were used safely and these, together with a growing awareness of the value of many insects, meant that farmers took a great deal more care when spraying.

The use of agricultural pesticides can create further problems for bee colonies

Because the insects that cause damage to crops either simply eat them or suck their sap, systemic insecticides seemed to be the way forward and these have been around for a very long time. The term 'systemic' applied to insecticides was coined in 1947 and the latest incarnations of these are the neonicotinoids that are now used, worldwide, on a vast scale and range of crops. These were developed in the 1980s and 1990s, include a number of different chemicals and are chemically related to nicotine. They were the first new class of insecticide in 50 years and proved very successful for a number of reasons: their toxicity for mammals is very low, they need to be applied only in very small amounts and their effects in the plant are long-lasting, although the dosage is calculated to give maximum protection during the early part of the plant's life and then to diminish as it is diluted by the plant's growth.

They can be sprayed onto the crop but are frequently used as seed dressings. As the seed germinates and the plant grows, the chemical is absorbed and taken into every part of the plant, including the pollen and nectar, where it is present in infinitesimally small amounts. Insects feeding on the plants or sap are killed due to the chemical acting on a particular pathway in the central nervous system. Bees exposed to 'high' doses of the chemicals, particularly clothianidin, are killed and this can happen where the dust from seeds being drilled is not controlled.

Other effects on bees are still being evaluated, including potential effects on the bees' ability to find their way home and, although there has been much discussion about the neonicotinoids, it must be remembered that there are many other chemicals also being used, all of which may pose a potential threat to bees.

The insecticide story is a fascinating and continually evolving one, but use of some sort of pest control will always remain a necessary part of agriculture and the risk to bees will remain. The skill is to balance the economic benefit with the risk.

As a postscript to this section, we must remember that the commonest chemicals in hives are those used by the beekeeper to combat pests of honey bees and that these, too, can cause problems.

STRESS ON BEES

The final section in this rather depressing chapter is a consideration of stress as applied to the bee colony. There are many factors causing stress, some of which the beekeeper can control and some of which (s)he cannot. Many of the problems detailed above contribute to stress: diseases and pests, poor/inadequate forage, chemicals, both in-hive treatments and agricultural pesticides, and anything which reduces the quantity and quality of available forage.

We must also include the weather, which has a major effect on plant growth, nectar production, colony growth, the ability of colonies to survive the winter and queen mating. The effects of weather are often ignored but it is a vital component in the equation.

Another vital component is the beekeeper: too much handling, poor handling, ignorance of the bees' needs, poor pest and disease control, careless use of chemicals in the hives, inadequate feeding when necessary and migratory beekeeping are all under the control of the beekeeper. Moving bees from crop to crop can cause an enormous amount of stress, both from the travelling and from the nutritional aspects. Bees need a mix of pollens and some plants provide pollen of poor quality so, where bees are restricted to one type of plant for a period of several weeks, this can have a large impact on the growth of the larvae and hence the colony. This is not so much a problem in the UK where acreages are smaller and other plants are usually available within flight distance at the same time, but it might be a major factor in countries like the USA and has probably been a factor in their problems with CCD.

This is not a complete list of all stress-inducing factors, but I do not want you, my reader, to become too depressed. We can generalise that stress is caused by anything which impacts on the natural life of the colony.

CHAPTER 8

BE POSITIVE

So what do our bees need and how can we help them? There are a few basic things that we can provide:

- a sound, clean hive
- adequate food at all times and particularly going into the winter
- a good supply of mixed pollens, particularly in spring and autumn
- good, careful handling as necessary for good husbandry
- a fertile, fecund queen heading the colony
- maintenance of colony strength
- absence of diseases
- low numbers (below 1000) of varroa at all times
- avoidance of chemicals as far as possible both inside and outside the hive.

All fairly obvious but essential if our bees are going to thrive. Many items in the list can be controlled, or at least influenced, by the beekeeper. Others cannot, but we should still be aware of them.

Although, all the bees in the colony are important, the winter bees, those hatching in August or later, are vital to the survival of the colony into the next season. If these are short of food or diseased, they will die early and the colony may not make it through the winter or it will be weak and may never catch up to produce a honey crop. The queen, too, is an essential part of the colony and, if she is poorly mated, diseased, or affected by chemicals, she can fail. The drones, which are essential for successful matings, are also vulnerable to varroa, acaricides and pesticides. There is no part of the colony which does not interact with all the other parts and, similarly, there is no part of the environment which can be separated from our bees.

Despite all the gloom and despondency, many colonies continue to thrive and this is frequently the result of good husbandry and adequate knowledge of the bees' needs. The job of the beekeeper is to reduce, as far as possible, the stress factors affecting his/her bees so that they can withstand those that are out of his/her control. Bees are remarkably robust creatures, have been around for a very long time and will continue to evolve to face the problems that affect them now and in the future.

9 FLOWER SHOW

The Angiosperms (Angiospermae, Anthophyta or Magnoliophyta, depending on which classification system you use) are the flowering plants. They evolved during the cretaceous period some 140 million years ago and became very widespread about 100 million years ago. They are the dominant plants throughout the world today and an enormously successful and diverse group, both in terms of the habitats in which they are found and the number of species.

The flowers which they bear are objects of beauty to all of us, a livelihood for some growers and florists, a larder for many insects and the source of beekeepers' honey, but for the plant they are reproductive structures and have evolved so that their parts are all directed towards ensuring that seed is produced to give a new generation of plants.

Flowers are a fundamental part of a bee's life and an integral part of beekeeping. Bees get all their food for their young, and themselves, from flower products and this is what sets them apart from wasps.

The best way to find out how flowers are made is to dismember them – they do not scream – and use a hand lens to study the different parts.

A section through an Apple flower (*Malus domestica*) showing five stigmas and styles and the ovary containing the ovules

HOW FLOWERS ARE MADE

To understand how a flower works we will look at the apple (*Malus* sp.) which is a member of the family *Rosaceae*, a large and varied family. The diagram and photographs should help you.

Flower structure

Stigmas, styles and ovary

These make up the female part of the flower and are found at the centre. The apple has five stigmas and five styles. The styles are

Transverse section through an apple showing the five-chambered ovary with two seeds in each part

The stamens surround the stigmas and styles

merely 'stalks' to hold the stigmas in a place where they can receive pollen. The ovary (the wall of which later becomes the apple core) is divided into five parts and each part contains two ovules. Each ovule contains a female sex cell (gamete), as well as other structures, and the ovules later develop into seeds. The whole collection of stigma, style and ovary may be called the pistil or the *gynaecium*.

Stamens

These are the male part of the flower. In the apple there are many stamens. Some of them form a circle closely surrounding the female structures and called the inner whorl. The rest surround this inner whorl and, not unreasonably, are called the outer whorl. Each stamen is divided into a two-lobed anther carried by a 'stalk' called the filament. The pollen grains develop inside the anther lobes and each pollen grain contains a male sex 'cell' (gamete) which is simply a nucleus. Collectively, the stamens may be called the *androecium*. Pollen is used as a food reward for some visiting insects.

Petals (5)

Their function is to attract insects, such as bees, to the flower, so they are large and colourful. Petals are collectively called the *corolla*.

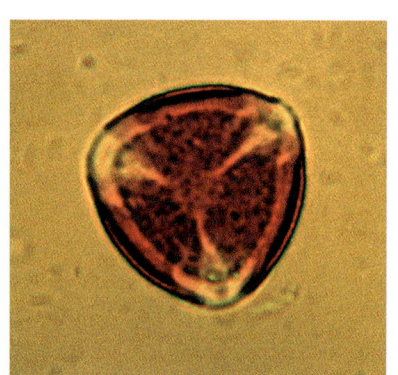

A pollen grain from Apple (*Malus domestica*)

The complete flower of *Malus domestica* is an object of beauty
The petals attract insects

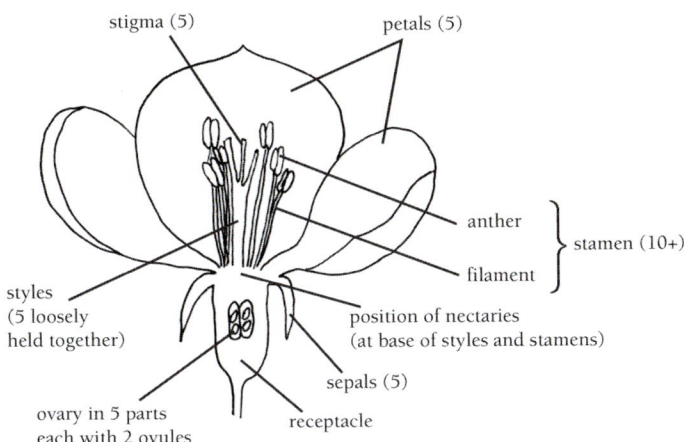

Section though a flower of Apple (*Malus* sp.)

Sepals (5)

These are like little green leaves. They surround and protect the flower when in bud and persist in the fruit as the shrivelled structure at the opposite end to the stalk. The sepals are collectively called the *calyx*.

Receptacle (sometimes called the torus)

All the structures so far described are attached to the receptacle (or, in the case of the ovary, contained within it). Nectar is produced by its cells and collects on its surface, providing a reward for visiting insects. More of that later. The receptacle later swells to produce the fleshy part of the apple, which we eat.

HOW FLOWERS WORK

There are two essential, but quite separate, processes which go on in a flower and result in the production of a seed:

1. *Pollination* is defined as the transfer of pollen from an anther to a receptive stigma.
2. *Fertilisation* is the fusion of a male gamete (from the pollen) with a female gamete (in the ovule) to produce a single cell called a *zygote*.

The process of fertilisation follows after pollination providing that the pollen grain is accepted by the stigma. The pollen grain develops

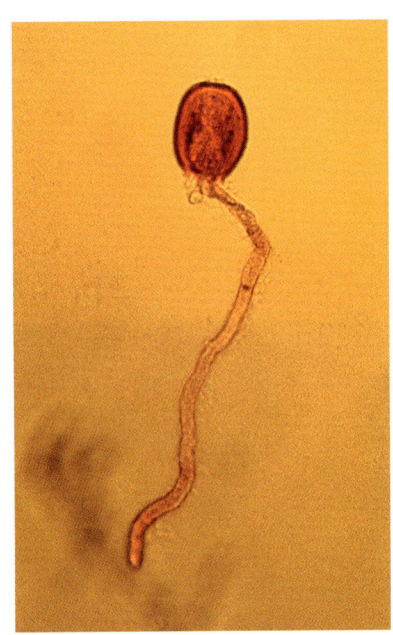

A pollen grain (of *Hippeastrum* sp.) showing the pollen tube which grows into the ovary through the stigma and style

A section through an Apple flower after fertilisation
Note the enlarged receptacle, the ovary and the developing seeds. The shrivelled remains of the flower are above

Longitudinal section of a mature apple
The flesh is the receptacle. The withered remains of the flower can be seen at the bottom

a tube, extruded through one of the tiny holes in its surface, which grows down through the stigma and style using enzymes to digest the tissues. When the tube enters the ovary, it locates an ovule and grows into it through a tiny hole called the micropyle. The male sex cell, carried in the tip of the tube, fuses with the female sex cell in the ovule and the process of fertilisation is complete. The resultant zygote will develop into an embryo plant and the whole of the fertilised ovule becomes a seed. Where there are several ovules, several pollen grains will be needed, one for each.

There is a slight complication in this process because there are two nuclei in the pollen tube (produced by the division of one original nucleus) which pass into the ovule. One, as we have seen, fuses with the female sex cell, but the other fuses with another nucleus in the ovule, called the primary endosperm nucleus, and this produces a structure which divides and develops into a food store, called the endosperm, which nourishes the developing embryo plant. The large, starchy part of a grain of wheat is an example of endosperm tissue.

Types of pollination

The definition of pollination given above suggests a simple process, but life is never that straightforward and there are two kinds of pollination:

- *self-pollination* where pollen from the same plant, usually the same flower, is used
- *cross-pollination* where pollen from a different plant, but of the same species, is used.

Self-pollination is easy to understand. Once the anther is ripe it bursts open, releasing pollen grains, which are received by the ripe stigma. Many plants use this method, including oilseed rape.

To achieve cross-pollination requires the movement of pollen to another plant which may be some distance away and to do this, *pollinating agents* are used. There are many pollinating agents worldwide including water, beautiful little birds, bats and other mammals, but in the UK most pollen is transported either by air currents (**anemophily**) or by insects (**entomophily**) and flowers have evolved so that they are adapted to either wind or insect pollination.

Two kinds of flowers

Briefly, insect-pollinated (entomophilous) flowers will usually be brightly coloured and conspicuous, may produce nectar and/or a scent and their pollen grains will be large and rough-coated or sticky. Wind-pollinated (anemophilous) flowers will have reduced, inconspicuous petals and sepals, anthers and stigmas which hang outside the flowers and stigmas which are often feathery and sticky. They will have no smell or nectar and the pollen grains will be small, light and smooth and be produced in enormous numbers. A word of caution here: some flowers (eg, willows, *Salix* spp.) have features associated with wind pollination but also produce nectar and are visited by insects. This mixed type of pollination is called *ambophily* and, while it is possible that it is a stable state utilising the best of both worlds, it may also be that these plants are in an intermediate stage of evolution.

Ways of ensuring cross-pollination

Elaborate methods have evolved in the plant kingdom to ensure that pollen from a different plant is used, at least some of the time:

- *self-incompatibility*. This is very common and means that, even if pollen from the same flower lands on the stigma, the pollen tube will not be able to develop and no fertilisation will occur. This characteristic is controlled by one pair of genes and an example is apple. It is well known that to give a good crop of apples, an apple tree needs another apple tree, of a different variety, but flowering at the same time. Both trees will provide pollen for the other. The need for another variety, in this case, is because apple trees of one variety, due to the way they are propagated, are all clones and therefore genetically identical. (A few apple varieties are self-fertile.)
- *protogyny and protandry*. Big words, but a simple concept. A stigma ripens when its surface becomes receptive to pollen. An anther is not 'ripe' until it has split open and released its pollen grains. So, in some plants such as Common Figwort, (*Scrophularia nodosum*), the stigma develops before the stamens (protogyny) or, more commonly, the stamens ripen before the stigma (protandry), as in Rosebay willowherb (*Chamerion angustifolium*).
- *heterostyly*. Here the lengths of styles and stamens differ in different plants so that pollen lodged on one part of an insect is deposited on the style at the same height in another flower, eg, Primrose (*Primula vulgaris*). Primrose flowers are of two types, pin-eyed and thrum-eyed, always borne

The flowers of Rosebay willowherb (*Chamerion angustifolium*) are protandrous

The two types of Primrose (*Primula vulgaris*) flowers
Thrum-eyed on the left, pin-eyed on the right

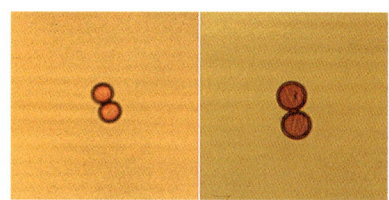

Pin-eyed Primroses have smaller pollen grains (left) than thrum-eyed flowers (right)
(Both at the same magnification)

Male catkins of Hazel (*Corylus arvensis*) shedding pollen

on different plants. This difference alone is not sufficient to ensure cross-pollination, but the pollen of the two types of flower is also different. That from the thrum-eyed flowers is bigger than the pollen from the pin-eyed flowers while the stigma of the pin-eyed type has projections on it which 'fit' the pollen from the thrum-eyed plant and vice versa. (Purple loosestrife, *Lythrum salicaria*, has three different kinds of flower with three different sizes of pollen grains.)

- *monoecious plants*. Separate male and female flowers are found on the same plant, eg, Hazel (*Corylus avellana*) The flowers may ripen at different times and, in any case, self-pollination is less likely when the two sexes are separated spatially.
- *dioeceous plants*. Plants may be of different sexes, eg, Sallow (*Salix* sp.) and Holly (*Ilex aquifolium*). This situation mimics that in most animals and self-pollination is clearly impossible.

Often more than one of these processes is involved in the same flower, and self-incompatibility is frequently involved, but, as the flower ages, this may break down and self-pollination will occur if no cross-pollination has taken place.

Why is cross-pollination important?

Cross-pollination must be advantageous to a plant otherwise all these mechanisms to ensure it occurs would not have developed.

When gametes fuse at fertilisation, each contributes its genes to the zygote. Where the gametes are from the same individual the resultant offspring will usually exhibit uniform characteristics, but where the gametes come from different individuals there will be more *variation* in the offspring because of the combination of different genes. Variation is the raw material for *evolution*.

Amongst the variable offspring, some individuals will be better adapted to their environment, or better able to adapt to a changing environment, than others. Such individuals will fare better than their fellows and, in this way, a species may gradually change and develop and new species will evolve. Without genetic variation, evolution is impossible apart from chance changes due to mutations of genes. In addition to this, cross-pollination will lead to the production of more vigorous offspring (hybrid vigour) and plants also tend to set more seed.

On the other hand, a flower is about producing seed so, if cross-pollination fails or is impossible, for whatever reason, it is better for the plant to use its own pollen than to produce no seed at all, so many plants use self-pollination as an insurance policy and some use it as a normal method, dispensing with the services of other plants and pollinating agents as a result. Often self-pollination takes over as the flower ages so that in apple, for example, which normally needs a pollinator, a few apples will usually be produced even if the tree is in complete isolation and unable to receive pollen from another variety.

The male catkins of *Salix* spp. are borne on 'male' trees

Female catkins of *Salix* spp. are borne on 'female' trees

Female flowers of Hazel (*Corylus* sp.) are carried on the same tree as the male flowers and pollinated by air currents

CHAPTER 9

White Clover (*Trifolium repens*)
The lower florets have been pollinated

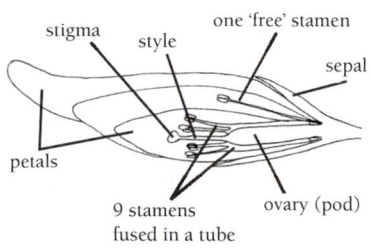

Section through a floret of *Trifolium repens*

Dandelion (*Taraxacum officinale*) can be an important nectar and pollen provider for a short time in spring

VARIATION IN STRUCTURE AND FUNCTION IN DIFFERENT PLANTS

The apple has a fairly straightforward flower but there is great variation and we are going to look at a few more flowers in detail. All these species are important nectar-producing plants and very important to us beekeepers.

White clover (*Trifolium repens*)

This is a member of the Fabaceae (formerly the Leguminosae). The common name for this family is the pea family and all of the plants have characteristic flowers.

- *Petals* (5). These are modified so that one large petal forms the standard, two are wings and two are fused to form the keel which encloses the reproductive structures.
- *Stamens* (10). Nine are fused to form a tube around the ovary, one is free and may be outside the keel. The nectaries are situated at the base of the stamen tube and nectar rises up the tube.
- *Stigma, style, and ovary*. There is one stigma and one style and the ovary is a long structure which later develops into a pod.

Another member of the Fabaceae is the broad, or field, bean (*Vicia faba*).

Dandelion (*Taraxacum officinale*)

The dandelion belongs to the family Asteraceae (formerly Compositae). In this family the individual, very small flowers are called *florets* and are built into a specialised inflorescence called a *capitulum*. This is surrounded by small green bracts. The capitulum in dandelion is flat and opens gradually, with the oldest florets round the outside.

Unlike some members of the Asteraceae, all florets in the dandelion have the same structure. The calyx is reduced to a number of hair-like structures called a pappus. (This will later form the 'parachute' which carries the seed when it is blown away.) The petals are fused into a corolla tube at the base but open into a flat structure above and the filaments are fixed to the upper end of the corolla tube. The anthers form a tube completely encircling the style. The ovary contains only one ovule and the nectary is around the base of the style. Dandelion flowers are odd because, although they exhibit all the structures and functions associated with other insect-pollinated flowers – colourful petals, nectar and large, rough pollen grains – their ovules develop without fertilisation and pollen is therefore completely superfluous and all the offspring from a particular plant are genetically identical 'clones'. This type of reproduction, which looks like normal sexual reproduction but has no fertilisation is called *apomixis*. It seems a backwards step in evolution but dandelions seem to do very well using this system, as everyone with a lawn will know.

Another example from the Asteraceae is Sunflower (*Helianthus annuus*) which has a different structure, possessing two kinds of florets, ray florets (like the dandelion) round the outside of the capitulum and disc florets in the middle.

A Dandelion 'flower' is an inflorescence made up of many florets

Lime (*Tilia* sp.)

There are several different species of lime which grow in the UK, the commonest being *Tilia x europaea* (syn. *T. vulgaris*), the Common Lime, *T. platyphyllos*, the Large-leafed Lime and *T. cordata*, the Small-leafed Lime, but we can take all the flowers as alike, although the size of the flowers and the inflorescence may vary from one species to another. The genus is the only one in the family Tiliaceae, but it contains many species.

The flower structure is based on the number five, so there are five sepals, five petals, numerous stamens roughly gathered into five bundles, one ovary, style and stigma but with the ovary divided into five sections. The nectaries are found inside the sepals, which are lined with hairs. These hairs hold the nectar and, because the flowers hang down, the nectar is well protected from loss due to rain and wind. Lime flowers also have a perfume. They are interesting because they exhibit features associated with both wind and insect pollination. They produce a perfume to attract insects and nectar to reward them but the flowers are not very conspicuous and the anthers and stigma hang outside the petals and sepals.

It is worth mentioning that not all limes seem to secrete nectar under all conditions and, where trees are being planted with bees in mind, it is worth carrying out some research to ascertain which are the best species and varieties in a particular area.

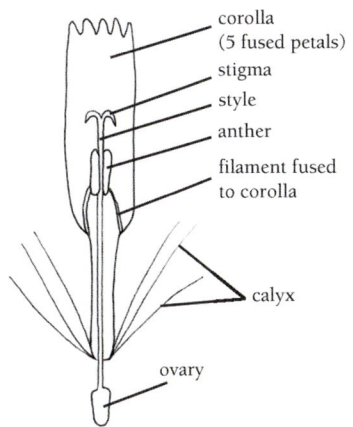

A section through a single floret of *Taraxacum officinalis*

Lime (*Tilia* sp.) flower with some petals and sepals removed to show the reproductive parts

CHAPTER 9

Heather (*Calluna vulgaris*)

Heather (*Calluna vulgaris*) grows in vast acreages on moorlands and also in rocky places

This is common heather or ling. It is the only British species in the genus Calluna and is the plant which covers huge areas of moorland in many parts of northern Britain, the West Country and Wales. It produces abundant nectar, under the right conditions, and the bees convert it into a dark, strong, thixotropic honey.

The flowers are pale purple in colour. They are built on the number four.

- *Sepals* – four, larger than petals but the same colour. (The sepals are described as 'petaloid' because they look like petals.)
- *Petals* – four fused into a tube in the lower part of the flower.
- *Stamens* – eight. The anthers are fused into a circle round the style and each carries two appendages or awns. The filaments are not joined together and are fixed at the base of the ovary.
- The *ovary* is made from four carpels (constituent parts) but there is one style and one stigma.

There are eight nectaries which are found between the bases of the filaments.

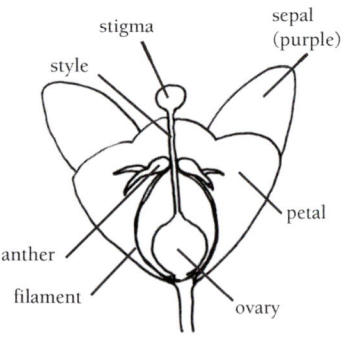

Section through a flower of *Calluna vulgaris*

Bell heather (*Erica cinerea*)

Most of the plants belonging to the genus Erica are called 'heaths' but *E. cinerea* is not! It flowers a little earlier than ling heather and produces a beautiful honey which is a rich port-wine colour and is not thixotropic. The flower is red/purple in colour, is similar to that of *C. vulgaris* and is also based on the number four. There are minor structural differences:

- The flower is more bell-shaped than that of *C. vulgaris* – hence the name.
- The sepals are smaller than the petals and are green.
- The petals are fused into a tube along a greater part of their length so that the flower is more enclosed than that of *C. vulgaris*.

Bee plants par excellence

Bell heather (*Erica cinerea*)

Both heather and bell heather are very valuable bee plants and beekeepers will often move hives many miles to obtain the honey from them. Like some other good bee plants they are not totally dependent on bees for pollination and much of their pollen is moved about by air currents.

The genus Erica is very useful and cultivated varieties have been developed so that the flowering season can extend throughout the year. Some of the early flowering varieties are well worth growing in the garden as early sources of nectar and pollen for our honey bees and for emerging queen bumblebees. They require little attention except for trimming once they have finished flowering so they will not take up valuable beekeeping time. But remember that they will only grow well on acid soils and do not like alkaline ones.

Common Sage (*Salvia officinalis*) is beautifully adapted for its insect visitors

Sage (*Salvia officinalis*)

Some Salvias are pollinated by birds but, in Britain, the native species use insects to carry their pollen. They are wonderfully adapted for their insect visitors and are among some of the most complex flowers. They belong to the family Lamiaceae. No it's not a misprint but the modern name for the family Labiatae (the dead nettles). Common sage is the one which goes with roast duck and which many people have in their gardens.

If you look at a sage flower you will see that the sepals and petals form a tube. There are really five sepals although there appear to be only two. This is because they are fused into a top lip with three teeth and a lower lip with two teeth (each 'tooth' represents a sepal). The five petals, too, are difficult to sort out. There is an

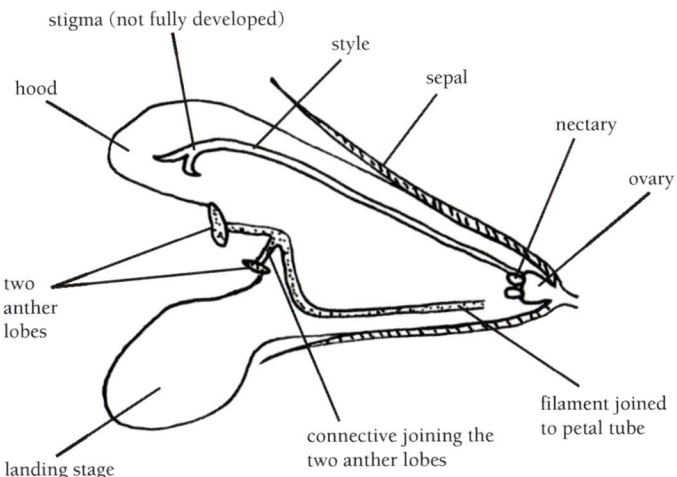

Section through a flower of Sage (*Salvia officinalis*)

upper hood made up from two petals and a lower 'landing stage' made up of three sections. The flowers are carried in a spike.

The female parts of the flower are straightforward – a round ovary at the base of the flower with a long style pressed under the hood. At the end is a stigma divided into two parts (known as a bifid stigma if you want to impress!). When fully developed, the stigma extends beyond the end of the hood. The ovary has four ovules in it and the nectar is found around it.

The male parts are the really interesting ones. The family Lamiaceae typically has four stamens but in Salvia there are only two. Those two are not simple. The anthers are two-lobed and these two lobes are connected by an extension of the filament called the *connective*. This exists in the anthers of other flowers and is normally very small but in Salvia it is extended so that the two lobes of the anthers are completely separated by two small 'stalks' attached to the top of the filament. The upper anther lobe is found under the hood and the lower lobes hang down into the opening of the flower. In *S. officinalis* both of the anther lobes produce pollen although the lower one does not have as much as the upper one. In many other species of Salvia the lower anther lobe is completely sterile.

The pollination mechanism

Because Salvias are such complicated flowers, they are usually pollinated by intelligent insects such as bees. The insect lands on the landing stage and pushes into the flower. As it does this it makes contact with the two lower lobes of the anthers which are pushed forwards. The connective is structured so that the other (fertile) anther lobe swings down onto the back of the insect's abdomen, so ensuring that it gets a good dusting of pollen. The flower is protandrous (the anthers develop first and then the stigma ripens later) so at a later stage the stigma extends past the end of the hood and comes into contact with the insect's back as it probes the flower.

In the natural world a partnership should benefit both partners and both the bees and the salvias gain from this one. Cross pollination, and seed production, is assured, so the flower can save on pollen production and the complicated flowers are beyond less intelligent insects, so the nectar is available for the brighter visitors instead of being taken by all and sundry. Both gain.

10 FOOD OF THE GODS

NECTAR

Bees' basic energy food is nectar and it is therefore the raw material upon which the beekeeper bases his activities, but why do flowers produce nectar? Plants may have begun to produce nectar as a method of controlling the concentration of their plant sap. Plant sap is the fluid that flows through the plant, carrying substances which the plant needs. In any biological system it is important to keep the concentrations of solutions inside the organism within fairly constant parameters. If they fluctuate widely, the chemical and physical processes going on inside the organism can be upset. In an actively growing, healthy plant, sugar may be produced at a faster rate then the plant can utilise it and it may need to remove some of it. Conversely, another view is that the nectar is a means of removing excess water from the plant.

There are structures found on plants called **hydathodes**. These are found at the edges of leaves at the ends of the tubes carrying water through the plant and allow water to escape when there is a danger of the plant having too much, such as in the first rush of growth in the spring when the roots may push a great deal of water into the plant. This can easily be seen in a plant such as the common nasturtium where water droplets collect round the edge of the leaves, usually overnight. I see it every year on my young tomato plants. The actual process is called **guttation**. It may be that some hydathodes have evolved to produce nectar rather than just water. Once a plant produces something which is attractive to an animal, whose visits benefit the plant in some way, there is a survival advantage if that co-operation is fully exploited. So, the production of nectar becomes a positive advantage and plants evolve to produce more nectar, better nectar or protected nectar. The flowers producing most nectar or the best nectar would attract more pollinators, resulting in more efficient seed production and a greater number of offspring from that plant. So the question of why plants produce nectar does not have a simple or obvious answer after all but, because of its basic relationship to beekeeping, its composition, where and how it is produced and what influences its production are of prime importance to all of us, although, as with many other facets of beekeeping, there may be nothing we can do to change things!

What is nectar?

Nectar is water (30–90%) containing a number of dissolved substances. These include:

- sugars
- vitamins, principally vitamin C and some of the B vitamins
- nitrogenous compounds such as amino acids
- minerals
- organic acids
- pigments
- aromatic compounds
- enzymes from the plant
- occasionally lipids or alkaloids.

The sugars, which are of primary importance to the bees and the beekeeper, are mainly sucrose, fructose and glucose (between them making up 5–70%) although several others have been isolated. Nectar comes from plant sap and the main sugar found in sap is sucrose but this may be broken down to glucose and fructose by plant enzymes. Nectars contain different proportions of the three main sugars. For example, most of those of the Brassicaceae (formerly Cruciferae) contain no sucrose and *Brassica napus* (oilseed rape) nectaries yield more glucose than fructose. Similarly nectar from *Taraxacum officinalis* (dandelion), a member of the Asteraceae, is high in glucose. Most of the Fabaceae nectars contain equal quantities of all three sugars, although *Trifolium pratense* (red clover) and *Robinia pseudoacacia* (false acacia) produce more fructose than glucose. Some of the cultivated rhododendrons give nectar with almost 100% sucrose. As we will see later, the sugar content of nectar can influence the physical properties of the resulting honey.

The remaining constituents of nectar make up about 3% of the total weight. Nectar should not therefore be considered an important source of vitamins, minerals or protein but the pigments and aromatic compounds are important to beekeepers because, although they are present in such very small proportions in nectar, they are concentrated in honey and are responsible for the colour, aroma and flavour of different honeys. Lipids (fats) and alkaloids occur in a few nectars and usually together.

How and where is nectar produced?

In my dictionary, a nectary is defined as 'a glandular organ that secretes nectar'. That about sums it up – it is a collection of cells which actively extract substances, mainly sugars, from plant sap and pass them to the outside of the plant. Most nectaries are found in flowers but they may occur elsewhere on the plant, when they are called 'extra-floral nectaries' – more of these later.

The nectaries are usually well-defined structures which take many forms such as slightly sunken or flat discs, protruding structures or slits or lumps, although it is often impossible to see them with the naked eye. Often they are surrounded by hairs and they can be found almost anywhere: on any part of a flower, but most frequently at its base. For example:

- in apple, and its relatives, the nectaries are found on the top of the receptacle
- white clover has its nectaries at the base of, and inside, the tube of stamens
- lime flowers have their nectaries on the inside of the sepals.

Typical cells in a nectary have thin walls, large nuclei and an abundance of cytoplasm and stomata (small pores) and often the walls of the outer ones are thickened, probably as a means of retaining the secreted nectar. They draw water and dissolved substances in and then these are exuded, either through one or more small holes or by oozing through the membranes of the cells. It is possible that some compounds are reabsorbed back into the cells surrounding the nectary, but this is certainly not definite. The structure of the nectary and its surroundings usually results in the nectar collecting around the nectary, sometimes in quite large amounts. (The largest amounts of nectar are produced by bird-pollinated flowers, many of them South American, and an example which you may be able to see at home is the Poinsettia, *Euphorbia pulcherrinum*.)

Structures within the flower may be modified to act as reservoirs for the nectar. The hairs on the sepals of lime flowers trap the nectar, the staminal tubes in clover collect it. The petals of *Helleborus* spp. not only secrete the nectar but are modified into 'pitchers' to store it. They are much appreciated by bees during the early part of the year. In other flowers, for example many of the Apiaceae (formerly

A flower of Oilseed Rape (*Brassica napus* ssp. *oleifera*)
A nectary can be seen at the base of the ovary. There are another three, of which two are partly visible

The flower of Poinsettia (*Euphorbia pulcherrimum*) is much reduced but has a large nectary, here containing a drop of nectar
The bird pollinators are attracted by the large colourful bracts surrounding the flowers

The Lenten Lily (*Helliborus orientalis*) has green pitcher-shaped petals where nectar is produced and stored. The purple structures are sepals

Umbelliferae) and ivy (*Hedera helix*), where the nectar is secreted at the top of the ovary of each tiny flower, there is no reservoir and the nectar simply sits there, unprotected and available to all.

The important point about the position of the nectar is that, in order to gain access to it, the visiting insect must come into contact with the stamens and stigma so that pollen exchange can occur. The whole subject of flower structure, nectar secretion and storage is so complex and fascinating that it is essential to read a good botany book and to dissect flowers in order to understand it. 'Dissect' sounds grand but this is something that, with the help of a good hand lens, a pair of forceps, notebook and pencil, anyone can do lying in the sun on a summer's day. And, if your nearest and dearest queries your apparent inactivity, you can loftily tell them that you are carrying out an important part of your studies!

Factors influencing nectar secretion

In general, a flower's nectaries produce nectar, which is taken by a visiting insect. It then produces some more and the length of time it takes to do that is very variable. Once the flower has been pollinated, nectar production will cease but other flowers, on the same inflorescence or elsewhere on the plant, will carry on producing nectar. The rate of production is not constant and there are many ways in which the environment, and the condition of the plant, will influence it.

Temperature

Some plants such as Blackberry (*Rubus fruticosus*) are not fussy but others need high temperatures. An example is False Acacia or Locust (*Robinia pseudoacacia*) and this is the reason why this tree provides a honey crop in warm countries but is generally useless in the UK. As a novice beekeeper, I had high hopes of the two False Acacias in the garden opposite mine but I was very disappointed. In southern Britain it can produce crops if the weather is co-operative. Many plants secrete better when nights are cool and days are warm or hot but lime (*Tilia* spp.) requires warm nights and warm, sultry days for best results. Willows (*Salix* spp.) and White Clover (*Trifolium repens*) are most productive at cooler temperatures although it is doubtful if it ever gets too hot in the UK to cause problems.

Time of day

As a general rule, nectar is often weaker during the very early part of the day and peaks during the morning as wind and heat evaporate some of the water. Although nectar quality may peak at one time, nectar quantity may peak at quite another and bees will have to balance the two to make their visits most profitable. Bees will often work one plant in the morning and then switch to another species later. I see this in my own garden where *Persicaria amplexicaule* is worked by huge numbers of bees in the first part of the day, but during the afternoon is scarcely visited at all. Conversely, Apple (*Malus* spp.) is usually worked by bees for nectar later in the day, although they may collect pollen from the flowers in the morning.

Wind and humidity

A light wind and low humidity are beneficial because these conditions can concentrate nectar, which has already been produced, making it more 'sugary' and therefore attractive to bees, but a strong, drying wind and very low humidity may cause production to cease or may dry the nectar to such an extent that bees cannot collect it.

Persicaria amplexicaule (on the right) yields nectar in the morning

Soil moisture

This is vital. If it drops to a level where plants wilt, secretion of nectar stops completely. Where soil moisture levels are low, deeper rooted plants, such as trees, will fare better than those with shallow roots. Soil moisture is a long-term matter, usually dependent upon the weather conditions months before the plant flowers. Rainfall during the winter is all important in building up soil moisture.

Nature of soil and subsoil

Acidity and alkalinity can influence nectar production just as they can influence which plants grow where. Acidity is measured on the pH scale where 7.0 is neutral, anything below is acidic and anything above is alkaline. Many soils in the UK are in the range 6.0–6.5, that is, slightly acid. The nature of the top soil is largely due to the rocks beneath it, so, for example, there is a characteristic flora in chalk and limestone areas and white clover yields best on these soils, which are alkaline. Heather (*Calluna vulgaris*) needs an acid soil to grow and to produce abundant nectar. Other plants such as Blackberry do equally well on acid or alkaline soil.

Age and vigour of the plant and position of the flower on the inflorescence

Generally, young plants in good health, growing actively, yield most nectar. This is noticeable on grouse moors where the heather is

White clover (*T. repens*) needs alkaline soil and moderate temperatures to give good nectar supplies

burnt off in rotation, the younger patches being more productive than the old, woody plants. Flowers lower in an inflorescence tend to produce more nectar. Watch the Rosebay willowherb (*Chamerion angustifolium*) which bees work avidly when it starts to flower but virtually ignore once the lower flowers have faded.

Topography

Heather and white clover both yield better on slopes rather than flat ground but this may be influenced by other factors, such as drainage. Frost pockets may cause problems with many plants.

Shading

Regardless of temperature, some plants yield well only in sunshine. The ice plant (*Sedum spectabile*) demonstrates this well. It is a very attractive garden plant, providing nectar for many insects, but where it grows in the shade, it attracts little interest from insect visitors.

There is no doubt that some plants are fussy, others are not and the factors listed above interact with one another. Some plants are just plain fickle and hawthorn (*Crataegus monogyna* and *C. laevigata*) is the one which springs to mind. Read the books and they will tell you that it yields one year in seven or needs this condition or that. I am surrounded by hawthorn, but, despite what the books say, I have never had a crop from it and, walking around our lanes, I seldom see a bee on a hawthorn flower! Different weather

Hawthorn (*Crataegus* spp.) is a fickle nectar yielder

conditions make no difference. Perhaps it is the soil, maybe it is the variety of hawthorn. There must be a reason, or several, but I am still waiting for the right combination of conditions and every year, being the eternal optimist, I look forward to Maytime and then anxiously scan the bushes for the sight of a bee.

On the bright side, our lanes are also lined with many blackberry plants and these always yield nectar. In fact, as you will see from several mentions above, blackberry yields under most conditions, is not fussy about soil or temperature and is so deep-rooted that it can always reach any available moisture. Added to these advantages, there are several hundred varieties of wild bramble so that the blossoming period is spread over several months. In a survey, carried out a few years ago, looking at the pollen in honeys, blackberry was found to be almost universal in honeys from various parts of the UK.

Blackberry (*Rubus fruticosus*) has many varieties, yields nectar under a huge range of conditions and flowers over a long period

Unusual substances in nectar

We have seen that there is a range of substances present in nectar and, therefore, honey and, unfortunately, a few of these are unpleasant. So, we have some nectars/honeys which taste and smell horrible and a very few which, taken in large amounts, are poisonous. In the UK, the two commonest unpleasant honeys are Privet (*Ligustrum* spp.) and Common Ragwort (*Senecio jacobaea*). Both have a bitter taste. I do not know much about Privet and it is very unlikely to be a problem as, only very rarely, are bees likely to collect large quantities of its nectar. Even so, a fairly small amount can damage the flavour of other nectars mixed with it. On the other hand, Ragwort is becoming more widespread, is very attractive to bees and is likely to produce quantities of extractable honey, which smells horrible when it is fresh. If it is allowed to stand and granulate, the flavour improves and some beekeepers use it to blend with other, less flavoursome honeys. The plant contains several pyrrolizidine alkaloids which are responsible for the deaths of quite a few horses each year when they are fed on hay containing ragwort. They cause liver damage and appear to be cumulative in the animal. These alkaloids also get into nectar, and honey, but precise information on the effects of such honey and the stability of the compounds during storage is not available. Recent work has shed a little more light on this problem, but it does seem likely that huge amounts of honey would have to be consumed to do any damage. There are other plants producing poisonous nectars, many of them in the family Ericaceae. These include *Rhododendron* spp. and *Kalmia latifolia* (both grown in this country but apparently

Ragwort (*Senecio jacobaea*) is a common plant
It contains potentially dangerous alkaloids

causing no problem). My experience of *Rhododendron* is that most of the time bumblebees use it but honey bees are not very interested, although this can change if the conditions, and presence of other flowering plants, vary. *Kalmia* is unlikely to occur other than as an isolated garden specimen.

Some nectars are harmful or poisonous to the bees themselves, which seems a bit of an own goal on the plant's behalf. The best known are probably the limes. Native limes are fine but there are three imported species which can cause bees to fall unconscious and result in many deaths:

- *T. tomentosa* (Silver Lime)
- *T. petiolaris* (Silver pendent, or weeping, lime)
- *T. orbicularis*.

The offending chemicals are naturally-occurring benzodiazepines, and some native American people use the flowers of *T. tomentosa* as a sedative. I have seen this phenomenon once where hundreds of bumblebees lay dead and dying beneath a small lime tree in full bloom and it does seem to vary from one year to another. Bumblebees appear to be more badly affected – perhaps honey bees have got more sense!

EXTRA-FLORAL NECTARIES

These have always seemed to present a problem. They are nectaries found on parts of plants other than flowers, so they have no connection with pollination. If my earlier suggestion about the evolution of nectaries from hydathodes is correct, then there is no reason why they should be restricted to flowers. This point is reinforced when we see that some non-flowering plants have extra-floral nectaries and that, in flowering plants, extra-floral nectaries have been described in at least 2000 different species from more than 64 families. So, they are not obscure exceptions to the rule, but very common structures.

Bracken (*Pteridium aquilinum*) has nectaries where the branches join the main stem and these can be important sources of nectar for bees under some conditions. In Guelder Rose (*Viburnum opulus*) and species of Cherry (*Prunus* spp.) the petioles (leaf stalks) carry paired nectaries and in Laurel (*Laurus nobilis*) they are discs on the underside of the leaves near the midrib. The Broad (Field) Bean (*Vicia faba*) has extra-floral nectaries on small black areas on the stipules, little leaf-like structures on the stems, making these plants even more attractive to bees and other insects. Those are a

Although Bracken (*Pteridium aquilinum*) does not have flowers, it still produces nectar in the coloured patches on the stem

The extra-floral nectaries of Cherry (*Prunus* spp.) are found on the leaf stalk (petiole)

few examples but there are many more. You may go slightly potty looking for them!

The generally accepted theory for the development of extra-floral nectaries, apart from helping to maintain the balance of water and dissolved substances inside the plant, is as a protective device. Many of these nectaries are very attractive to ants and these can provide protection to an otherwise vulnerable plant. Ants will deter larvae, and bigger animals, which may defoliate the plant. Certainly it is very easy to watch ants getting nectar from broad bean extra-floral nectaries and one experiment demonstrated that, when broad beans had a proportion of their leaves removed to simulate damage by herbivores, the plant produced a lot more extra-floral nectaries within one week. This does seem to suggest a defence mechanism and I am thinking of educating my cabbages in the benefits of extra-floral nectaries!

Most nectar is produced from these nectaries when the plant is growing actively in the early part of the year, which is when there is abundant sugar available and also when most damage can be done to the developing plants.

Laurel (*Laurus nobilis*) has extra-floral nectaries on the undersides of its leaves, either side of the midrib

RECYCLING – BEE STYLE

Honeybees never make life simple do they? Just when we think we have learned all about the source of our honey, a complication looms ahead in the shape of honeydew.

The small black areas on the stipules of Broad Bean (*Vicia faba*) are extra-floral nectaries

Origins

One of the orders of Insects is the Hemiptera, which includes all the bugs. A huge group, it is divided into two smaller groups, or sub-orders, and one of these is called the Homoptera. This group includes lots of very small but quite troublesome insects. Many of them are important pests of plants (a few suck blood, but they do not worry us in this context) and include the aphids, scale insects, coccids and many more. They are beautifully adapted to their way of life, with mouthparts which are modified into a system of piercing 'needles' and tubes for transporting liquids.

The higher plants have, within their stems and leaves, tubes which transport dissolved food materials around the plant. These are called the phloem. Phloem carries what is usually referred to as plant sap and it is this which most members of the Homoptera 'tap into' to feed.

Another split second and the drop of honeydew being excreted by the aphid will be flicked away

How honeydew is produced

When an aphid (or other Homopteran) pierces the phloem with its mouthparts two things happen:

- a duct within the mouthparts pours saliva into the wound, starting the digestion process
- a second duct transports plant sap up to the insect.

The plant transport system is interesting and important in understanding the production of honeydew. The liquid solution moves through the phloem under pressure and when the tiny insect mouth parts tap into it, the liquid is pushed into the insect without it having to make any effort. If you watch a feeding aphid (a good hand lens is necessary), you will see that, once plugged in to this liquid food supply, the insect stays put. You may also see that, at intervals, a drop of moisture appears at its rear end and is shot, by a flick of the insect's body, away from it. This is honeydew. I have spent many hours photographing aphids and have frequently observed honeydew being excreted. The aphid raises its abdomen and usually waves it around for a short while. The drop of honeydew is suddenly shot away by movements of either the abdomen or the hind legs. Its production may be rapid under good conditions and, where the sucking insects are present in large numbers, as aphids usually are, there may be a large amount of honeydew produced, which collects on the leaves.

Honeydew is not plant sap

So, tiny insects take in large quantities of plant sap, extract what they need as it passes through the gut, and excrete the remainder through the anus. They probably need to do this so that they can obtain sufficient of the various nutrients, most likely vitamins and minerals, which occur in minute amounts in the sap. But the stuff that comes out is not identical to what went in! On its passage through the insect's gut it has been changed to some extent.

Composition of honeydews

- mainly water.
- up to 90% of dry matter is sugar.
- the range of sugars is large and includes, in addition to glucose, fructose and sucrose, complex ones such as

melezitose and fructomalose. Some of these are synthesised inside the insect.
- nitrogenous substances make up 0.2–1.8% of the dry matter. Most of this is amino acids but there is usually some protein. Some of these substances are synthesised by the insect.
- there are always organic acids, such as citric acid, present.
- there are various enzymes present. These come from the saliva which the insect pours into the wound in the plant. Others are added as the liquid passes through the gut.
- there are usually moulds and spores present. These can be seen under the microscope and act as a means of identification for honeydew honey. They come from the moulds which move into the honeydew as it lies, exposed, on the surface of the leaf. (Where leaves are covered by honeydew they often go black due to the presence of sooty moulds.)

Aphids, such as this Black Bean Aphid (*Aphis fabae*) produce live young at a rapid rate
The resultant colony can excrete a great deal of honeydew

A feast for all

Honeydew is used as a food source by insects other than bees. I have watched wasps and flies taking it and ants are notorious for feeding on it, often forming a column of marching insects leading to an aphid colony. Because it is produced in small drops and is exposed to the air, it dries quickly. Bees often collect it in the morning when there is a dew to liquefy it.

Some people regard the honey made from honeydew as a luxury; others will not touch it. That is purely a matter of taste. It ranges in colour from light brown to almost black and may have a greenish tinge. Its flavour is strong and it is slow to granulate. In the Black Forest region of Germany, the honey produced from honeydew collected from the aphids on the pine trees is prized

I look on it as yet another example of interaction between insects but with a less violent outcome than most.

11 OTHER BEE ESSENTIALS

Honey bees have five basic requirements around which their lives are built and so far the only one we have considered is nectar. In this chapter, we are going to look at the other four: pollen, propolis, water and the only one that the bees manufacture, beeswax.

SMALL AND PERFECTLY FORMED – POLLEN

'Small', in this context, is an understatement because pollen grains are microscopic in size and 'perfectly formed' applies not just to the external appearance, which is often beautiful, but also to the structure of the pollen grain, which, as we have seen in Chapter 9, is perfectly suited to its function. A pollen grain is packaging for the essential nuclei inside it. They are surrounded by an envelope of cytoplasm but all of this has to be protected, sometimes in very difficult conditions, so a pollen grain is surrounded by a thick wall which is divided into two parts:

- the *intine* which is tough but not hard
- the *exine* which is very hard and often sculpted into amazing shapes. It has a number of small openings, or apertures, in it which provide a means of escape for the pollen tubes.

The exine can survive all sorts of harsh treatment and appears to be almost indestructible so that the little, vitally important nuclei within the grain are delivered safely to their destination, the stigma of a flower. This ability to withstand different conditions is the main reason that the flowering plants have become such a successful group, as the pollen grain has released them from the need for water, during reproduction, which is necessary in more primitive plants such as mosses and ferns. This simple fact has enabled them to colonise a wide variety of habitats, many of them comparatively

A bee with well-laden pollen baskets

hostile. Because of its indestructibility, the exine is the part of the pollen grain which is characteristic of it and enables us to identify it. But why would you want to? There are a number of reasons:

Identifying the source of honey and determining whether it is what it says it is. For example, by identifying pollen grains it is sometimes possible to say that a honey labelled as 'English' is imported, or at least contains some honey from elsewhere. Analysis of the pollen content of some so-called 'monofloral' honeys can demonstrate that they are nothing of the sort. Prosecutions have been brought successfully using pollen analysis as evidence. The technique of sampling and identifying the pollen grains in honey is called *melissopalynology*. The pollen is obtained by diluting 10 g of honey with 20 ml of hot water. This is mixed and divided between two tubes which are centrifuged for 10 minutes. The contents of the two tubes are then mixed, the liquid poured off and the tube filled with water. This is then centrifuged for a further five minutes. The pollen grains will be thrown to the bottom of the tube and can be drawn off with a pipette. If you do not have a centrifuge about your person, all is not lost but you need a little patience. The honey can be well mixed with 100 ml of water and left to stand overnight. The pollen will settle to the bottom. You may have to repeat this more than once.

Flowering plants became the dominant plant group during the Cretaceous Period between 145 and 65 million years ago, so pollen analysis is used in archaeology as a method of dating various finds and deposits and understanding some human activities.

In forensic science, pollen is sometimes used to help in fixing the location of a crime or a suspect's whereabouts. Exciting stuff this. Apparently all woodlands have a characteristic pollen 'fingerprint' and this has been the undoing of several criminals, including murderers. So, if you are going to murder someone, keep away from incriminating plants and be careful where you dispose of the body!

Identifying pollen grains

If you want to study pollen grains you need a compound microscope. A magnification of x400 (four hundred times) is necessary (a x10 eyepiece used with a x40 objective will produce this). For more detailed work, and for smaller grains, a x100 oil immersion objective is very useful (giving a magnification of x1000 when used with the x10 eyepiece). An eyepiece containing a graticule (which is a short subdivided scale visible when you look through the eyepiece) allows accurate measurement of the pollen grains. Probably the only way to get to grips with the subject is to build up a collection of microscope slides of pollen grains, collected directly from flowers, so that you can compare the pollen from other sources with this reference set.

Preparing a slide of fresh pollen collected from a flower seems complicated until you try it. It is really very simple and, once you have done a few, it takes very little time. All you are doing is staining the pollen so that its detail is more visible and mounting it so that it is permanently available. To do that you have to remove the oil which often covers pollen grains, stain and mount the grains, cover with a coverslip and seal with a waterproof sealant. Then do not forget the most important bit – the label. You will need:

- microscope slides and coverslips.
- a warming plate. You can buy these but I use a good Heath Robinson type made from a small tin with a very low power electric light bulb inside it.
- ether or iso propyl alcohol.
- stained glycerine jelly (glycerine jelly to which a small amount of basic fuchsin has been added, so making it pink). Purists use clear glycerine jelly and add their own stain, either fuchsin or safranin, so that they can vary the amount they use. Fuchsin and safranin both stain the grains pink.
- a means of warming a small bottle. Warm water in a bowl does the job.
- a waterproof sealant to seal round the coverslip once the slide is dry.
- a sticky label which is fixed to the slide.

How to proceed

First get the pollen you wish to mount. If you are using pollen from flowers, it is best to pick the flowers before they start shedding pollen when they are just on the point of opening. Bring them indoors and keep them warm for a few hours, preferably overnight. You can then obtain the pollen by tapping/shaking it onto a microscope slide. Alternatively, the pollen may have come from a pollen load taken from a bee. In this case, put the load onto a microscope slide with a drop of water and use a glass rod, or something similar, to gently break up the load and mix it with the water. Remove most of the sample, leaving a small amount in the middle of the slide. Then allow it to dry. Similarly, if it has come from a honey sample, draw up some of the pollen sediment, put it onto the microscope slide and allow the liquid to dry. All samples can then be treated in the same way:

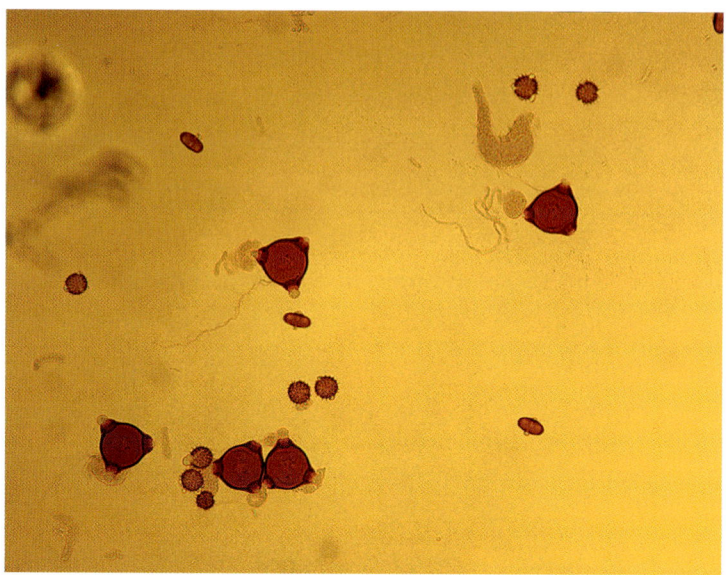

Mixed pollen of Rosebay willowherb (*C. angustifolium*), the large ones; Sunflower (*H. annuus*), round and spiky; and Hogweed (*H. sphondylium*), oblong/oval

- add a little of the alcohol or ether to the pollen. This removes the oil from the surface and washes it. Some pollens are more greasy than others and may need 2 washes.
- remove the residue of alcohol/ether and allow the rest to dry by warming it (on the warming plate).
- meanwhile warm the glycerine jelly (in the water bath) so that it is runny.
- put a drop of the jelly onto the dried pollen on the warm slide.
- lower the coverslip, *gently*, onto the sample from one side, being careful to exclude air bubbles. The jelly should just occupy the coverslip completely without any squeezing out round the sides.
- when dry, seal round the edge of the slide with a sealant to keep out moisture. Sealants can be purchased but nail varnish or paraffin wax also works.
- label the slide.

Producing professional-looking pollen slides comes with practice. You will soon find the exact amount of jelly to put on the slide and will learn to keep your hand steady while sealing. Like many newly learned skills, it requires patience, but even slides which are not perfect from the aesthetic point of view can be useful as part of a reference collection, so do not be discouraged.

What are you looking for?

Having prepared the slides, the next job is to study them. The characteristics used to distinguish between pollen grains are straightforward:

1. How big is it? Sizes are measured in microns (μm). There are 1000 microns in 1 millimetre so they are small. Pollen grains vary between 7 μm and 100+ μm with many around 20–30 μm.
2. What shape is it? It may be round, rectangular, somewhere between the two, triangular, irregular, etc.
3. How many holes (apertures) does it have? You remember these are the openings in the tough exine. They can vary in type and are called furrows if they are long and thin and pores when they are just small holes.
4. What is the exine like? This outer, very hard covering is varied in appearance. It may be smooth, look as if it is covered by a net, be marked by lines, carry bumps or spines, and so on.
5. How thick is the exine? You can see this by focusing the slide up and down carefully. Often there are several layers of the exine visible, with quite distinctive characteristics.
6. There may be other characteristics not covered above such as the presence of air sacs. Pollen grains may be stuck together in groups of four, called tetrads. The apertures may have thickened edges or caps over them. The intine (inside layer) may be very thick.

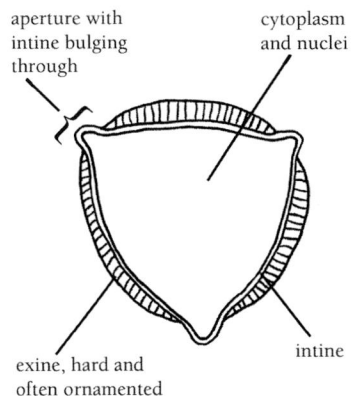

A pollen grain

This is not a complete list but it gives you an idea of the factors involved in identification of a pollen grain. Some species are very easy to identify and others are extremely difficult, although it is often possible to pin it down to a family of plants. To take the subject further, there are several books about pollen and these are listed in the bibliography.

Identifying pollen loads

Looking at the pollen loads brought in by our bees is a fascinating part of beekeeping life and a different aspect of pollen identification. It is often possible to tell in this way which flowers the bees are working because the pollen loads have characteristic colours. Of course, a little common sense has to be applied because the majority of pollen loads are in the colour range yellow to orange. So you must know what flowers are out in your area. If you have lots of time to spare (and who has in the hectic summer months?)

Beetles such as these Soldier Beetles (*Rhagonycha fulva*) are the most primitive group of insect pollinators

A Hover Fly (*Eristalis tenax*) feeds on a Dandelion

Members of the Lepidoptera are good pollinators, have long tongues, here seen partly coiled, and are very hairy
This Elephant Hawkmoth (Deilephila elpenor) frequently feeds on Honeysuckle (Lonicera periclymenum)

you can watch the bees collecting pollen from flowers and see the colour of the load but failing that, there are books to help you.

The importance of pollen to insects

As we saw in Chapter 9, insects are employed by many plants to transport their pollen from one plant to another. Only four groups of insects are important pollinators. These are the beetles (Coleoptera), the two-winged flies (Diptera), the butterflies and moths (Lepidoptera) and the bees, which belong to the Hymenoptera. As flowering plants and pollinating insects have evolved, the two groups have developed close relationships and, in return for acting as go-between for the plants, insects usually receive food rewards. One of these rewards is pollen. This may seem a little strange when the plant needs its pollen for reproduction, but plants having an efficient means of moving pollen about can afford to make much more pollen than they need and allow the insect visitors to steal some. For our honey bees pollen is vital, providing the nutrients, particularly proteins, for various needs:

- the large amounts needed by young worker bees making the food fed to brood, the queen and young drones
- the pollen mixed in small amounts with brood food and fed to older larvae
- the protein and lipid (fat) content of the fat bodies of worker bees necessary for overwintering
- the building materials for eggs and sperms
- the smaller amounts needed by the adult bees to produce enzymes and hormones and to enable other minor body repair and building jobs to be done.

In addition to the protein content of pollen, approximately 20%, there are small amounts, about 5%, of lipids (fats) and some minerals and vitamins are also present. It is estimated that a colony of honey bees uses about 20 kg of pollen per year but the other insect groups listed above, as pollinators, do not have this great need for pollen. Their early stages feed in different ways and, although many of the adult Coleoptera eat pollen, the Diptera, Lepidoptera and the non-bee members of the Hymenoptera visit flowers almost exclusively for nectar. Because bees depend upon flowers, totally, at all stages of the life cycle, their value as pollinators is immense and the fact that honey bees have such large, perennial colonies to support, and therefore have a need for a great deal of food, makes them pollinators *par excellence*.

Honey bees have developed special pollen baskets (*corbiculae*) on the tibiae of their hind legs and these enable them to carry quite large loads of pollen back to the nest. The pollen is packed into the pollen baskets as the bee travels from flower to flower and the average load carried back to the nest is 15 mg.

STICKY STUFF

As well as collecting nectar and pollen from plants, honey bees also use some of them to provide *propolis* or bee glue. This is the name given to the substances after they have been collected by the bee. Propolis is often mentioned almost as an afterthought but is an essential part of bee life and is an extraordinary substance, although a bee colony probably only uses about 100 g in a season.

Worker bee carrying a load of propolis in its corbicula

What is it?

It is the collected exudates of some plants, particularly trees such as poplars where it is found on the buds, and conifers, where the bark often exudes large quantities of resin. There are many other plants, including sunflowers, which yield resinous substances and, because the sources of propolis are so varied, it is not possible to give accurate details of its composition. However, it does contain:

- 50–55% resins and balsams
- 10% essential and aromatic oils.

These probably need a little explanation. They are organic compounds. The resins are transparent or translucent substances with quite long molecules, a feature giving them flexibility and the ability to soften when heated. An example is shellac. The balsams are similar and are aromatic. Examples are Canada balsam (used in microscopy, but now largely superseded by other products) and frankincense, of biblical fame. Many resins and balsams are used commercially for various purposes.

The essential and aromatic oils are so-called to distinguish them from the fatty oils. They are often sweet smelling and are commercially important. They include *terpenes* and here we are on more familiar territory as examples are pinene, found in most of the essential oils in the conifers, camphor and geraniol (an important constituent of Nasonov pheromone). Many of these are used in perfumery and for other purposes. In addition to these two groups of substances, propolis contains *waxes, various acids,*

Propolis used to seal a crack near the entrance of a hive
The individual loads, varying in colour, can be seen

esters, a mix of *flavonoids* and other compounds, some of these added by the bees. Flavonoids are one of the most characteristic classes of compounds found in flowering plants and often function as plant pigments.

Many of the substances in propolis, particularly the flavonoids, have disinfectant properties and also act against bacteria, fungi and even viruses, so propolis is extremely valuable to the bee colony.

We think of propolis as sticky, and, at the temperature of the brood box, around 35 °C, it is, but as the temperature falls it becomes harder, finally becoming brittle at 5 °C. Propolis can range in colour from pale yellow through various shades of brown to red/brown.

How do bees use it?

The clues are in the properties of propolis, both its physical properties of stickiness, flexibility and hardness at low temperatures and its ability to combat micro-organisms. So we can construct a list of uses:

- filling cracks and crevices, so preventing draughts and helping to control some enemies. For example, wax moths are prevented from getting into crevices to lay their eggs.
- reducing entrances both permanently, in many wild colonies, and particularly during the winter, in hives. Some strains of the more hardy bees construct curtains of propolis just behind the entrance to keep out wind and rain; others fill the whole entrance leaving only single bee openings in the wall. This is where the word propolis comes from: *pro* means 'before' and *polis* means the 'city' in Greek.
- covering the inside of a natural cavity such as a hollow tree. This helps to keep the cavity clean and dry and keeps the walls smooth, preventing loose bits of wood from breaking off, and effectively provides an antiseptic envelope around the nest.
- it is mixed with beeswax to strengthen the base of the combs where they are fixed to the wall or roof of the nest.
- cells are varnished with propolis before the queen lays in them. This helps to combat disease-causing organisms.
- any intruder, such as a slug, which dies or is killed but is too big for the bees to remove, is completely covered in propolis. This prevents it from rotting and causing problems in the nest.

Human uses

Humans have found uses for propolis, particularly in the medical realm. In ancient times it was used as a dressing for wounds and this is being investigated again today. It is used in some medicines, such as treatments for sore throats. An enormous number of claims have been made for it including activity against tumours and respiratory infections. It appears to have anti-viral and anti-bacterial activity and is even emerging as a potential tool against HIV. It is also claimed that it reduces hair loss but the method of application is not detailed and I leave that to the imagination. We know that sometimes it induces allergies, usually resulting in a type of dermatitis, and I personally know two beekeepers who have become allergic to propolis after many years of trouble-free beekeeping.

In ancient times it was used in embalming and it has been a constituent of varnishes.

Propolis and beekeeping

To beekeepers, propolis is a nuisance – it clogs up everything, making it difficult to separate hive parts and frames, and it covers the hands or gloves with its sticky brown stains so that everything becomes difficult to handle and the nearest and dearest complains about sticky steering wheels, gear levers and suchlike. This is another example of conflicting interests between bees and beekeepers. It is difficult to remove. Hot washing soda will do the job but you do not really want to wash your hands in that too often. Alcohol will also dissolve it and is usually effective at cleaning up the hands. On equipment, a smear of petroleum jelly will prevent the bees from propolising a surface. I used to have a number of Manley frames which suffered very badly, becoming glued together in one solid super-full, but, on the few occasions when I actually remembered to take the petroleum jelly with me and use it, the problem was solved.

Some strains of bee use a lot more propolis than others so, if you are very troubled, it might be worth changing your bees. The amount that is collected also depends to some extent on availability. If, like me, you live surrounded by trees including many pine trees, you really do not stand a chance but, as more research work has been done, it is becoming clear that propolis is a very important tool in the bees' immune system, conveying a certain amount of protection from disease-causing organisms, including viruses. This makes it invaluable to the bee community and it may be that bees that use a larger amount of propolis are healthier than those that use it very sparingly. Perhaps selecting for bees that use little propolis is not such a good idea.

There is a market for propolis if you can find it. You can collect it on purpose-made screens, chip it off when it is cold and sell it, but you will have to ensure that it is free of other substances such as wax and bits of wood and, unless you are very dedicated, you might feel that life is too short.

WATER

Why do bees need water and how do they get it? They carry it back to the hive in their crops in exactly the same way as they carry nectar. Bees often seem to prefer water containing organic matter, so they will choose muddy puddles rather than sparkling sources. Usually only a small number of bees collect water but their numbers can be rapidly increased if the need arises, particularly in very hot weather. We can list the various reasons for water use:

- to dilute honey stores ready for feeding. Remember that nectar is the bees' natural food and honey is concentrated nectar.
- to dissolve granulated honey.
- for manufacture of brood food by the young bees. Brood food is about 70% water.
- to cool down the interior of the nest if it threatens to overheat. The water is spread on the comb, and elsewhere, and evaporated by fanning bees.

A bee carries water back to the hive in its crop

BEESWAX

In this section I am hoping to turn us all into chemists, because we are going to look at the composition and properties of beeswax. I approach the whole subject with some trepidation because beeswax is a very complicated substance and the available information tends to be confusing. Add the usual quota of long names and you can see that we may be in for trouble!

There are some facts which must be stated at the outset:

- all insects produce wax and it is an integral part of the insect cuticle, providing a waterproof layer
- the composition of wax varies in different insects
- bees produce more wax than other insects and its composition is not constant.

OTHER BEE ESSENTIALS

How it behaves

The properties of beeswax are reasonably straightforward:

- it is a white solid when newly produced (in fact the individual flakes of wax are colourless).
- it melts at 62–64 °C.
- it becomes workable at 32–35 °C.
- its Relative Density (Specific Gravity) is approximately 0.96.
- its acid number (or acid value) is 20. This means that it takes 20 mg of potassium hydroxide to neutralise 1 g of wax.
- some of its components react with alkalis to form a soap. (Hard water contains alkaline calcium salts which will react with beeswax to form a white deposit on the surface of the wax, so it should never be used to wash beeswax.)

A brief chemistry lesson

In most bee books there is a list of the constituents of beeswax with no explanation of what they are, so I will attempt to explain a little further.

Organic chemistry is concerned with substances based on the element carbon (C). The substances formed are called organic compounds, are vast in number and make up all living things. (As beekeepers we are very familiar with many of them, eg, sugars, proteins, thymol, formic acid.) The thousands of organic substances are arranged in groups. Each group is called a series and the substances within each series have similar chemical properties and the same general formula. This latter point may seem a little strange until we see that the carbon atoms upon which the substances are based can be built into chains. So, in any series of organic substances, the simplest member will have one or two carbon atoms and the most complex may have, theoretically, any number.

Within the molecule of an organic substance, the atoms are held together by bonds which are 'shared' between adjacent atoms. Each carbon atom forms four bonds. Because the hydrogen (H) atom can form only one bond, four hydrogen atoms can form bonds with one carbon atom and the substance produced is the simplest organic compound, methane, which has the formula CH_4. This can be shown in a simple diagram:

$$\begin{array}{c} H \\ | \\ H-C-H \\ | \\ H \end{array}$$

If one of the hydrogen atoms is removed and a second carbon atom and its attendant hydrogen atoms is attached, we have the next substance in the series, ethane, with the formula H_3C-CH_3 where – is the bond connecting the two carbon atoms. We normally write the formula CH_3-CH_3 for reasons which you will see in a moment because, if we do the same thing again we create the third member of the series, propane: $CH_3-CH_2-CH_3$, and so on. The members of this series are alkanes (or paraffins) and alkanes belong to a large group containing several series called *hydrocarbons*, which means they contain carbon and hydrogen only. Classification of organic compounds is really rather similar to the classification of animals that we looked at in Chapter 1.

Exchanges and action units

Alkanes are very stable compounds because all their bonds are satisfied, but one or more hydrogen atoms can be taken away. Because the carbon atom always needs four bonds, a double bond can be created and this is then a member of a second series called the alkenes (= is a double bond):

CH_3-CH_3 can become $CH_2=CH_2$

Ethane   Ethene (ethylene)

The double bond is not so stable as a single one. Ethene is the first of the alkene series and notice that these substances are also hydrocarbons. You should be getting the idea by now! Also, one or more of the hydrogen atoms can be replaced by another group of atoms altogether (O is oxygen and always has two bonds):

CH_4 can become CH_3-OH

Methane   Methanol (methyl alcohol)

or

CH_3-CH_3 can become CH_3-COOH

Ethane   Ethanoic acid (acetic acid)

In the group –COOH, one oxygen atom is connected to its carbon atom by a double bond and the other is connected to it by a single bond and also connected to the hydrogen atom by a single bond. Methanol is the first in the series of alcohols and ethanoic acid is second in the series of organic acids.

We can show their structures like this:

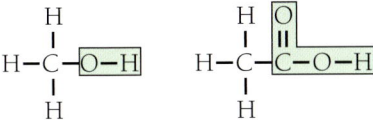

Methanol   Ethanoic acid

The units which replace the hydrogen atoms are the parts of the molecule which enter into reactions and can therefore be called *action units*. So –OH and –COOH are action units.

I have chosen these two series because when an acid and an alcohol react together they produce a substance called an *ester*, plus water. Using our two examples above:

$$CH_3-COOH + CH_3-OH \longrightarrow CH_3-COO-CH_3 + H_2O$$

Ethanoic acid + Methanol   Methyl ethanoate + Water

Methyl ethanoate is an *ester* (it is also called methyl acetate). Esters are very important substances in beeswax.

Components of beeswax

There are something like 300 different organic compounds in beeswax. Many of them belong to the series of substances that we have already discussed but they have long carbon chains. There is a very good reason for this. Generally speaking, substances with short chains are either gases or liquids. As the chains get longer, the compounds get more solid and, of course, beeswax is a solid material (although with a comparatively low melting point). A very rough list of the main categories of beeswax components would be:

- hydrocarbons   14%
- acids   12%
- esters   60% or more.

There are many other compounds present in small amounts and the proportions may vary.

The *hydrocarbons* include many alkanes with between 21 and 33 carbon atoms in their chains (you can amuse yourself writing out their structure if you've nothing better to do!). The major acid is *cerotic acid* with the formula $CH_3-(CH_2)_{24}-COOH$. At least the formula is something like that but there seems to be some disagreement and it is probable that it is a mixture of acids, all with very similar formulae.

There is little in the way of alcohols but **myricil alcohol**

[CH_3–$(CH_2)_{29}$–OH] is the main one and is important because the major ester is *myricil palmitate* which is derived from myricil alcohol and palmitic acid. There are several other esters of myricil alcohol also present.

Complexities

It is difficult to simplify a subject such as beeswax structure but I hope that, although quite complicated, explaining the chemical background will make it easier to remember the long names. My short potted account of organic chemistry is very basic indeed and I must qualify it by mentioning that carbon atoms do not always form straight chains but may form branched chains or rings. There are many different action units giving rise to many series of compounds. All organic substances contain carbon and hydrogen, many also contain oxygen and others may contain other substances such as chlorine, phosphorus and sulphur. It is a fascinating branch of chemistry which is well organised and logical. The only problem, as ever, is the long words and the fact that all the names have changed since I learned my organic chemistry (this is why I have often given two names), so if you are tempted to pursue the subject further, make sure that you get an up-to-date textbook.

Making beeswax

A worker bee exudes flakes of wax from the wax glands beneath her abdomen

The raw materials for all the substances in beeswax come from nectar/honey and pollen, although pollen probably contributes a very small part. As we have seen, the basic constituents for the various compounds in beeswax are carbon, hydrogen and oxygen and these are all found in the sugars in nectar.

To make large quantities of wax, the colony needs a good supply of incoming nectar and a temperature of 35 °C. The secreting bees hang together in clumps and wax production proceeds at a surprisingly rapid pace. The wax is produced by the four pairs of wax glands found on the underside of abdominal segments 4 to 7. These are thickened areas of the epidermis with many fat cells and *oenocytes* (cells particularly involved in the production of lipoproteins) lying over them. The cells of each gland thicken and develop during the early part of a bee's life and reach their peak of development when the bee is 16–18 days old. The haemolymph transports the raw materials needed for the manufacture of the wax and these are extracted by the oenocytes which produce the individual compounds needed for the wax. The constituent parts of the wax pass through spaces between the thickened cells of the wax gland and then move to the outside of the body through the miniscule canals in a sheet of very thin material lying below the gland and called the wax mirror. Here, the wax hardens and forms a tiny transparent flake which is removed by the back legs of the bee and then moulded and manipulated by the mandibles.

Use of beeswax

Honey bees use beeswax to make their homes. They mould it into wax cells which are built into sheets called combs. The cells are all the more miraculous because they have a constant shape and size. The bees rear all their young in these cells, store their food in them, make slightly larger ones in which to rear drones and construct special peanut-shaped cells in which they raise new queens. In the natural state, without the interference of beekeepers, the combs are usually built parallel to one another with curved lower edges and are spaced so that the bees can work between them. Adjacent combs are attached to each other by brace comb which strengthens the entire structure. The comb is versatile and strong, often needing to support a heavy weight of honey. The hexagonal shape is the most space-efficient, giving maximum usable cells with no spaces between them. It appears that the cells are built as cylinders and the wax flows, at the high temperature in the cluster, into the familiar hexagons. The cells are built back to back with a thin sheet of wax between them and walls offset to give added reinforcement. Very clever.

12 ON BEST BEHAVIOUR

How do your bees behave when they fly away into the big wide world beyond the hive? Well, I hope! But what, exactly, are they doing and why do they do it? Easy, you might say: they leave, fly to a flower and bring back nectar or pollen and that, on the face of it, is true but, as in all beekeeping things, life is never so simple. In this short chapter we are going to explore the foraging behaviour of our honey bees.

Early training

Worker honey bees generally work within the hive for the first three weeks of their lives and then they move outside to become foragers. This change is programmed into them by their hormonal system, changes in their glands and the development of parts of their brains, but it is variable and onset of foraging can be delayed, or accelerated, depending on conditions and need in the colony. As they change from house bees to foragers, major changes occur in their bodies: juvenile hormone levels rise and levels of vitellogenin, a very important protein, fall. It is these changes which trigger the move outside and also start the aging process leading to death in two to three weeks.

For some time before this they will have been leaving the hive for short exploratory flights. This, I think, is one of the loveliest sights in beekeeping – hundreds of young bees flying backwards and forwards in front of, and facing, their hives, fixing their positions and appearances in their brains. They also absorb information about the situation of their hive, landmarks and the patterns of polarised light in the sky, essential navigating knowledge for a bee far from home.

It is perhaps worth mentioning that some research workers believe that the lunch-time 'play' flights with which we are so familiar are not restricted to young bees, but may include older foragers waiting for the opportunity to accompany a queen on a mating flight.

Young worker bees take exploratory flights to learn the position of their hive before beginning to go out foraging, but are they all young?

CHAPTER 12

Choice of commodity

Once the bee is ready to start work as a forager, it leaves the hive and ventures forth to collect one of four commodities:

- nectar
- pollen
- propolis
- water.

Before it leaves the hive, a bee knows what it is going to collect – no impulse shopping here. The requirements of the hive will determine the nature of the trip. A few will collect water. They prefer water containing rotting vegetation and will not go far from the hive. Bees at a water source may use their Nasonov pheromone or, possibly, a footprint pheromone, to mark the area. A few may also collect propolis if it is a warm day. These bees often keep doing the same job as long as the weather is right, but the most likely activity of the forager is the collection of nectar, pollen or both and the bee will initially go about this in one of two ways:

- *either* it watches a dance by a returning forager and goes out to find the same forage
- *or* it leaves without watching a dance and searches until it finds some suitable flowers. Only a few do this and they become scouts, at least for a time.

Acting on information received

A dancing honey bee will 'tell' other bees 'watching' it where there is a good source of forage. During the dances, the bee will give samples of nectar, containing the scent of the flowers, to the watching bees who will also use their antennae to detect any floral scent on the dancing bee's body. The newly recruited forager will then leave the hive, fly in the direction, and for the distance, indicated by the dance and will locate the flowers using the remembered scent. On future visits it will use colour and shape of the flowers as well as scent. (You will notice that all these statements are made with a reasonable degree of confidence, but I have a sneaking suspicion that not all bees are as efficient or as intelligent as we assume and I wonder how many set out and never get there or go in the wrong direction. There must be stupid bees and bees with a poorly developed sense of direction or just plain idle bees!)

The language which bees use to communicate forage sources, is quite advanced. The direction of the source is always indicated by the angle made with the vertical and the vertical always represents the sun, wherever it is. The length of the 'waggle' run, the straight part of the dance, is closely correlated with the distance to the source and one second of the run roughly equates to one kilometre. It is vital to understand that the dances take place in darkness inside the hive so the waggle run, where the bee rapidly moves its abdomen from side to side, is accompanied by a buzzing sound made by the bee's wings.

The distance appears to be computed by using optical flow and work done with bees using patterned tunnels showed that a bee travelling through a heavily patterned tunnel danced accurately, whereas a bee travelling through a tunnel with stripes going along it did not and indicated a much shorter distance. Where part of a flight is over a lake, the distance indicated shows that bees do not register the distance over water. Where a source is close to the hive, the round dance is used to tell the watching bees to 'go out and look nearby'. The dancing bee does not do a figure-of-eight with a waggle, but simply goes round in a circle and then reverses and goes back round the circle in the opposite direction. Such information leads to many bees hunting around for forage and can cause problems such as robbing.

Learning and experience

Once a bee has found its flower it has to discover how the flower functions. Many are simple but if we take a flower such as white clover, the bee has to understand which parts to open to reach the goodies inside. It makes sense, therefore, for an individual bee, once it has learned a floral mechanism, to stick to it and there is evidence that this is what happens. It is called *crop constancy* and is more efficient both in terms of time and effort for the bee and pollination for the flowers, because pollen is not transported to unrelated species. Some bees carry this to extremes and may never move from a forage source even when it is failing. (We have all seen this: the few bees still investigating the one or two flowers left on a field of oilseed rape long after all the others have moved off to field beans next door.) Most bees will be recruited to new forage by dances in the hive as their current flowers become less attractive.

Sometimes bees are a bit too clever for the good of the flowers when they learn that bumblebees have made holes in the bases of flowers, such as runner beans. They are then able to 'rob' the flowers of their nectar without using the proper flower mechanism.

A honey bee uses the hole chewed in the base of a Broad Bean flower to 'rob' it of its nectar

In some cases, for example tobacco (*Nicotiana* spp.) this behaviour may enable the honey bee to reach nectar which would otherwise be unavailable to it because of the length of the flower tube.

Fresh fields and pastures new

A few bees are naturally more inquisitive and restless than others (the research workers of the bee world?) and do not exhibit crop constancy. They are always trying new flowers and, when they have found suitable new forage, they recruit other workers to it. These are the scouts and serve a vital function in fully exploiting the area around the hive. The numbers of scouts vary and their method of working is not understood. They probably happen upon new sources by chance as they sample many flowers. When there is good forage available, such as fields of borage for example, there will be fewer scouts than at times when flowers are scattered and not so productive. When there is a real dearth of nectar, most of the foraging bees may act as scouts.

Collection and transport of nectar and water

The honey bee has mouthparts adapted for sucking up liquids such as nectar and water. It has a long proboscis (tongue) made by the meshing together of a number of components. The inner part is called the glossa and within it is a tube. Surrounding the glossa is an outer tube called the food canal. Saliva pours down the cavity in the glossa from the glands above it and the outer food canal opens at the top into a cavity called the cibarium, which can expand because of muscles pulling on it, causing liquid to be drawn up the tube. From the cibarium, the nectar passes into the pharynx, which can also expand, and then into the oesophagus, which leads to the crop (honey stomach). This is a bag which expands to carry the liquid back to the hive. Theoretically, the crop can carry 100 mg, but in practice the average load is 40 mg. The far end of the crop is shut off from the rest of the digestive system by a valve, the proventriculus, which can open to let some food through.

Collection and transport of pollen

The proboscis/crop mechanism, though perfectly adapted for collecting and carrying nectar or water, is useless for carrying home solids, so the worker honey bee has been equipped with a whole set of tools to enable it to collect and transport pollen (and propolis). All of the hairs on a bee's body get covered with pollen or, if a bee is deliberately setting out to collect pollen, it can remove it from the anthers using its mandibles. The forelegs and middle legs are used to comb out the pollen from the hairs on the bee's head and thorax respectively. The inside of each basitarsus is covered with stiff hairs like a clothes' brush and this pollen is passed to the

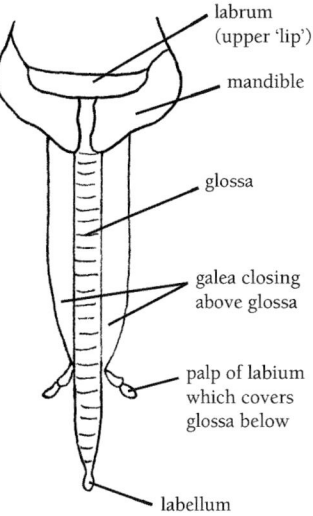

Mouthparts of a worker bee
(slightly separated)

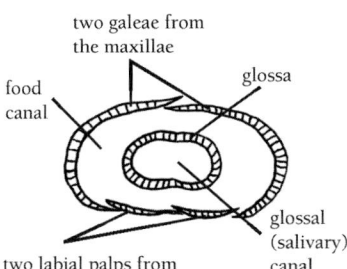

Transverse section through the proboscis of a worker bee

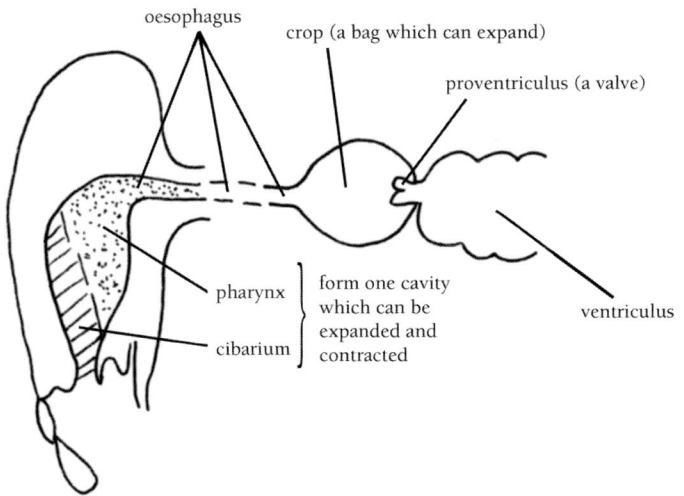

Nectar is drawn into the bee and transported in the crop (honey stomach)

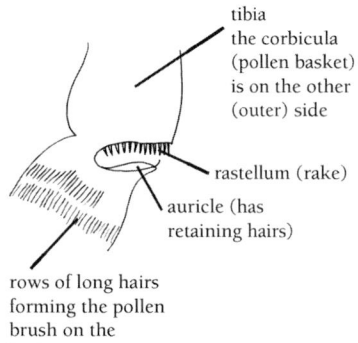

The inner side of the tibio-tarsal joint on the hind leg of a worker honey bee showing the pollen press

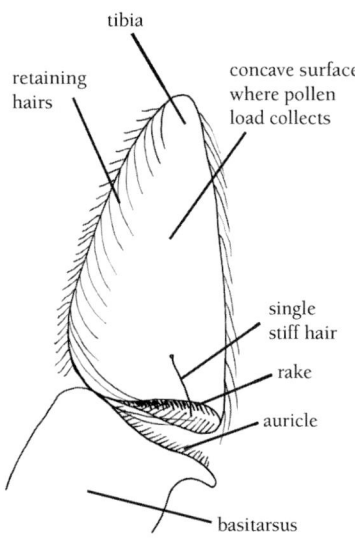

The outer side of the tibia of the hind leg

inside of the basitarsus of the hind leg, which has rows of bristles, called the pollen brush, and traps the pollen. The bee is than able, by rubbing its back legs together, to transfer the pollen from the basitarsus on one hind leg to the pollen press on the other hind leg. The *pollen press* is the name given to the structures found between the basitarsus and the tibia, making up the tibio-tarsal joint and is made up of:

- the *rastellum* (or rake) made of stiff hairs on the inside of the tibia
- the *auricle* which is a flattened hollow on the end of the basitarsus. It tilts upwards and outwards, is covered with small teeth and has a fringe of hairs on its outer edge.

As the bee combs through the opposing pollen brush with the rake, the pollen falls onto the auricle. The retaining hairs keep it there and, when the bee squeezes the joint together, the pollen is pushed upwards onto the outer surface of the tibia. This surface is concave, fringed by retaining hairs and called the pollen basket, or *corbicula*. In its centre is a long stiff hair which is essential to the task of building the pollen load. It takes the bee some time to build its pollen load and some bees are more efficient at it than others. It is possible to watch a bee hovering in front of a flower and packing pollen into its pollen baskets, although the movements are too rapid to see in detail. The pollen is moistened with a little nectar to help it stick together.

Once back at the hive, the bee drops its pollen loads into a cell where they are then processed by house bees who add secretions to them. These come from the hypopharyngeal or mandibular glands and from the bees' crops and a little honey is also added. The pollen undergoes a lactic acid fermentation brought about by bacteria from the bee's digestive system. This results in a more nutritional food, containing extra vitamins, and also ensures that the pollen will keep. The resultant mix is sometimes called 'bee bread' and is eaten in large amounts by very young adult bees as the main precursor of brood food.

The problem of propolis

Bees carry propolis back to the hive in their pollen baskets (corbiculae). Most is gathered in the late summer, or autumn, on warm days (when the propolis is soft and can be worked) and it is often possible to see a bee in the hive with two glistening drops of propolis on her legs. The loads are in the colour range of browns,

A bee disappears into the hive with a load of propolis

are shiny and easy to recognise. Collecting the loads is quite difficult and time consuming and bees occupied with this job do not collect anything else. A bee uses its mandibles to pull a small amount of the soft propolis away from the plant. It is then transferred, using the forelegs, to the basitarsus of the middle leg which presses it into the pollen basket on the hind leg on the same side of the bee's body. So the process does not involve the pollen brushes on the hind legs or transferring it from one side of the bee to the other, as is the case with pollen loads. Once back in the hive, the bee is unable to dislodge the loads so they have to be bitten off by other house bees. It is then used immediately and is never stored.

Statistics

Finally a few numbers for those of you so inclined:

- average nectar load carried by a bee is *40 mg* although they can carry considerably more
- average pollen load carried (both legs) *15 mg*
- number of nectar collecting trips per day is *10–15* on average
- number of pollen collecting trips per day is *7–13*
- percentage of mixed pollen loads (loads containing more than one type of pollen) is *1–10%*
- weight of propolis needed by a colony per year is approximately *100 g*.

13 MORE THAN SWEET, THICK AND STICKY

In earlier chapters we looked at nectar production in plants and the relationship between honey bees and plants and now, in this final chapter, we have to look at honey. In my dictionary, honey is defined as 'a sweet, thick fluid elaborated by bees from the nectar of flowers'. Fine! We all know that. Another dictionary includes the word 'sticky'. Even better! But we need a little more information on the nature of honey other than it being sweet, thick and sticky.

The Honey (England) Regulations 2003 give us a more precise definition: 'honey' means the natural sweet substance produced by *Apis mellifera* bees from the nectar of plants or from secretions of living parts of plants or excretions of plant-sucking insects on the living parts of plants which the bees collect, transform by combining with specific substances of their own, deposit, dehydrate, store and leave in honeycombs to ripen and mature.

That is quite a description! Honey bees produce honey as a long-lasting food for use in times of nectar shortage. After all, if you have many thousands of mouths to feed you do not want to be dependant upon the weather and flowers entirely, wherever you live. Man discovered honey at a very early stage and learned to rob honey bees' nests to obtain it (together with the brood which was an important source of protein and is still eaten by many people). In some parts of the world, this is still the principal method used to obtain this prized food but, man being always inventive, it was not long before it dawned on him that it made life easier if the bees were located somewhere close to his home, so that he could plunder their stores at will, rather than having to trek, maybe miles, to find them. So beekeeping was born.

At first it was primitive, as it still is in many parts of the world. The bees were housed in natural containers, such as hollow logs, or 'hives' were constructed from the materials available locally such as clay or straw. Whatever the method, the ultimate prize was the honey which, before the discovery of sugar, was the only sweetener available, provided a source of energy-rich food for people who had very little money or land and also could induce happy oblivion

Liquid honey

when fermented and turned into mead, probably the oldest alcoholic drink known to man.

Honey was endowed with many mystical qualities and, along with the bees which produced it, was present in the symbolism associated with most religions, ancient and modern. Even today it retains this air of wonder and mystery and many people swear by its ability to promote health and longevity, as well as to cure hay fever. Any beekeeper who has had much contact with the general public will endorse this. So to practicalities. Honey bees are invaluable as pollinators and, although that is their real importance to mankind, most beekeepers keep bees in the fervent hope that they will produce large crops of honey. We need, therefore, to know something about this substance, what it contains, what its properties are and how we should handle it to get the best possible product.

NECTAR TO HONEY

A physical change

As we saw in Chapter 10, nectar is a solution of sugars in water with a few other things thrown in. The foraging worker bee collects nectar and carries it back to the hive in her honey stomach (crop). When she gets home she gives it to another worker bee, or several, by regurgitating it onto her proboscis and passing it to the receiver bees who carry it away and put it in a cell in the honey storage area. There is usually more than one receiver bee taking the load and the nectar may not all reach the storage area because the receiver bees and others may eat it. A bee working in the storage area reduces the water content of the nectar by taking a drop onto her proboscis and rolling it up and down so that the drop is exposed to the high temperature in the hive (34.5 °C). As a result, water evaporates from the nectar.

This process is repeated many times by many bees, the nectar being returned to a cell after its treatment. The water passes into the hive atmosphere where the humidity is maintained at 40–50%. Surplus water must be drawn out of the hive by fanning bees, some lined up at the entrance, while others fan on the combs. The resultant air currents remove some of the moisture. This gives us one of the pleasures of beekeeping life – the steady hum set up by the vibrating wings of hundreds of bees in the evening, following a good day's nectar collection, removing excess water from the hive and, with it, the scent of the freshly-gathered nectar. Then we can dream of full supers.

MORE THAN SWEET, THICK AND STICKY

Once the water content is down to around 18%, the worker bees seal the cell with a wax capping so that air and, most importantly, water are excluded. The honey will then keep for a long time and beekeepers describe it as 'ripe'. It is perhaps worth pointing out the obvious, which is that nectar takes up a great deal more space than honey because it contains so much more water. In a practical sense, this means that supers must be added to enable bees to store the maximum amount of nectar, but those supers will not all be full once the nectar has been converted to honey.

Chemical changes

Nectar contains sugars, mainly sucrose, glucose and fructose. Glucose and fructose are called monosaccharides (simple sugars) because their molecules cannot be split into smaller units by the action of dilute acids. Incidentally, in older books glucose is called dextrose and fructose is called laevulose. Sucrose, on the other hand, is a disaccharide. Each of its molecules is made up of two molecules of monosaccharides linked together. One molecule is glucose and the other is fructose. Sucrose can be split into its two constituent parts by dilute acids. This reaction, which is reversible, can be represented by a simple equation:

$$C_{12}H_{22}O_{11} + H_2O \rightleftharpoons C_6H_{12}O_6 + C_6H_{12}O_6$$

$$\text{sucrose} + \text{water} \qquad \text{glucose} + \text{fructose}$$

The observant amongst you will see that glucose and fructose have the same chemical formula and this can be confusing. In fact, the two molecules are built differently giving them different properties. The process of adding water to the sucrose molecule, however it is done, is called hydrolysis.

There are trisaccharides, formed by three molecules of monosaccharides and so on, but these do not concern us, although they do occur in small quantities in some honeys. Honey bees convert most of the sucrose in nectar into its two constituent parts but they do not use acid. Instead they produce two enzymes which act on sugars.

- *sucrase (invertase)* which is able to do the same job as a dilute acid, in converting a molecule of sucrose to a molecule of glucose plus a molecule of fructose. The sucrase comes from the bee's hypopharyngeal glands and its production starts when the bee is about two weeks old, continuing until it dies. The foraging bee adds some sucrase to the nectar

when it collects it and the processing bees continue so that most of the sucrose is converted to glucose and fructose by the time the honey is ripe.
- *glucose oxidase* is also produced in the hypopharyngeal glands and is added to the nectar. This acts on some of the glucose, breaking it down into hydrogen peroxide and gluconic acid. The hydrogen peroxide is particularly important to us as it destroys bacteria.

Composition of honey

Honeys vary considerably in their composition – it would be very boring if they did not – but it is possible to give an average breakdown of honey constituents. The following figures have been taken from several sources:

Constituent	Amount as % of whole
Water	17–19
Fructose	38–40
Glucose	31–35
Sucrose	1–3
Other sugars	8
Total acids	0.5
Ash (mineral content)	0.09–0.33
Nitrogen	0.04–0.05

In case I have not given you sufficient numbers to remember there are two more:

- average pH of honey is 3.9
- average diastase value is 20.8 (see below).

I must stress that there is great variation in all these figures and that the ranges around these average figures are quite large. In particular, the variation in the higher sugar content can be enormous and can change with time since some of the higher sugars are formed gradually by chemical reactions going on in the extracted honey.

There is a large number of acids which may be found in honey. The major one is *gluconic acid*, produced, as we saw above, by the action of glucose oxidase on glucose. As well as contributing to the flavour of the honey, the acids are responsible for the low pH but some of the minerals present also affect this number.

Mention of minerals brings us to ash content since the minerals are what is left once the honey has been reduced to ash. I could produce a long list of metallic and non-metallic elements found in honey as research workers in many different countries have devoted a great deal of time to this topic, but there really is no point. Potassium is usually the largest component of the minerals but there are many more. Remember that we are talking about a very small constituent of honey – considerably less than 1%.

The nitrogen in honey comes from a number of different compounds including proteins and a small amount of amino acids. The amino acids come from the bee itself and the major one is proline.

The diastase number (activity) can be used as an indicator of overheated honey. Diastase (or amylase) is an enzyme which breaks down starch and it comes from the bee. Like all enzymes, it is destroyed by a high temperature and by long storage at lower temperatures. In the UK, honey must show a diastase activity of not less than 8.

The diastase activity is determined by measuring the rate at which starch is broken down. It is often stated in Schade units, 1 Schade being the amount of enzyme that will convert 0.01 g of starch into the prescribed endpoint in 1 hour at 40 °C. The test is called the Schade test but there are other methods of determining the activity. This is completely superfluous information but it is of interest and, I believe, helps the understanding of what appears to be an arbitrary number.

It is useful to consider the half-life of diastase in honey (that is, the time taken to reduce the diastase activity to half of its original number). At 20 °C it is about four years, at 30 °C about six-and-a-half months and at 70 °C about five hours. Those figures illustrate the value of storing honey in fairly cool conditions. There are some monofloral honeys which have naturally low diastase activity and these are taken into account when determining whether the honey has been overheated.

Although the water and sugars make up the greatest part, the minor constituents of honey are all-important from the point of view of aroma and flavour. There are also compounds formed, in small quantities, by the breakdown of other substances within the honey. Some of these occur over time and are dependent upon the treatment given to the honey once it leaves the hive. One breakdown product of importance is *hydroxymethylfurfural (HMF)*. This is formed by the breakdown of fructose in acid conditions and, in the UK, honey must not contain more than 40 mg/kg. As with diastase, HMF is formed gradually over a period of time at low temperatures, but its production is accelerated where the honey is subjected to high temperatures. Measurement of HMF therefore gives an indication of the degree to which honey has been heated and its storage conditions.

Some factors which can affect composition are:

- original source of the nectar – overwhelmingly the most important
- geographical location
- weather conditions
- soil type and mineral content of the rocks and soil
- actions of the beekeeper when extracting, processing and storing the honey.

HONEY BEHAVIOUR

We have seen how honey is produced and dealt with its chemical composition but honey, like any other substance, has certain physical properties which govern how it will behave and, in this section, I will to try to give a simple explanation of those properties and their implications. Actually it will have to be simple as I am no physicist!

1. *Density* is probably the most straightforward. It is defined as the ***mass per unit volume*** of a substance. It is stated in kg per cubic metre (kg/m^3) but in the good old days it was stated as lb per cubic foot and sometimes it is expressed as g per litre. The density of honey is approximately 1400 kg/m^3.
2. *Relative Density* which used to be called Specific Gravity (S.G.) is the ***ratio of the density of a substance at a given temperature (usually 20 °C) to the density of water at 4 °C***. Because it is a ratio it has no units but, because the density of water is 1000 kg/m^3 at 4 °C, numerically Density and Relative Density are the same. The relative density of honey at 20 °C. is therefore 1.4.
   Practical applications
   The density of honey varies with the honey type and this is important when honeys are being mixed or blended as honeys with different densities will form layers if they are not mixed very thoroughly when warm.
   Density and Relative Density also vary with water content. This variation has enabled the *hydrometer*, an instrument used for measuring relative density, to be used to measure the moisture content of honey and so determine its 'ripeness' and keeping qualities. Although in theory a hydrometer measures relative density, it can be calibrated to give readings of moisture content. A hydrometer must be used at the temperature stated on the instrument since relative density varies with temperature.
3. *Viscosity* is defined as the ***resistance to flow that a fluid offers***

MORE THAN SWEET, THICK AND STICKY

when it is subjected to sheer stress. Put simply, a fluid with a high viscosity will be 'thicker' than one with a low viscosity. Viscosity reduces rapidly as the temperature rises from 5 °C to room temperature. Above about 30 °C there is little change. Viscosity also reduces as water content increases.

Practical applications

Honey is often warmed to reduce its viscosity and enable it to pass through fine filters. In theory viscosity can be measured and used to determine moisture content of honey, but in practice it is not used except sometimes experimentally.

4 *Thixotropy* is the *rate of change of viscosity with time*. A thixotropic substance becomes less viscous when it is moved rapidly and the longer it is moved, the less viscous it becomes. We are familiar with the properties of ling (*Calluna vulgaris*) honey, which is thixotropic. By this we mean that when it is allowed to stand it becomes very viscous and jelly-like but when stirred rapidly it becomes fluid. This property enables ling (heather) honey to be extracted, usually with some difficulty, by a normal honey extractor providing that it is stirred vigorously first. In this instance, the increase in viscosity is due to the presence of proteins in the honey which alter its state. Despite its jelly-like nature, heather honey may have a high water content. Some other honeys such as manuka (*Leptospermum scoparium*) from New Zealand show the same property.

5 *Hygroscopicity* is a *tendency to absorb water*. For each honey there will be a relative humidity at which no moisture will be either lost or gained. This is called the equilibrium relative humidity. Relative humidity is the ratio, expressed as a percentage, of the moisture in the air to the moisture it would contain if it was saturated at the same temperature and pressure. Honey stored in contact with air will tend to absorb moisture from the air if the air is at a higher relative humidity than the equilibrium and lose water to the air if the relative humidity is lower than the equilibrium. The surface layer will absorb the water first and, as a result, will have a lower density than the lower layers. It will tend to stay on the surface because of the high viscosity of honey which reduces movement within it.

Practical application

Honey should be stored in full containers with well-fitted lids which exclude air and, as far as possible, at no stage in its extraction and processing should honey be exposed to high levels of humidity.

6 *Refraction* is the final property I want to mention. Light travels at different speeds in different substances and the changes in the wavelength cause the light to 'bend' as it passes from one

A refractometer is a useful tool

medium to another. We are all familiar with this phenomenon occurring at the surface of water and altering the appearance of an image. *Refractive Index* (or, more correctly, relative refractive index) is defined as the *ratio of the speed of light in a vacuum (or air which is pretty close) to the speed of light in the medium*. Refractive index varies with variation in temperature.

Practical application

A *refractometer* is an instrument used to measure refractive index. Because the refractive index of honey varies with moisture content, a honey refractometer, calibrated to give direct readings of water content, is commonly used. It is a convenient instrument, readily available and comparatively cheap. It is calibrated for use at a specific temperature.

The physical properties of honey is not a subject which usually engenders much enthusiasm among beekeepers but I hope this section has at least explained the basics and given a better understanding of why honey behaves as it does. Ideally, it should be thick and sticky and, perhaps, now we can see why. But we still have two very important aspects of honey behaviour to investigate.

GRANULATION

Most UK honeys granulate after a time

Freshly extracted honeys are liquid but, once they are left to stand, most honeys begin to form crystals and eventually become solid, frequently rock hard. This process is called granulation or crystallization and produces granulated/crystallised honey. It happens because, at lower temperatures, honey is a super-saturated solution of sugars. *Super-saturated* describes a solution holding more of the dissolved substances than would be dissolved if the system were in equilibrium. Adding a small crystal will induce the solution to crystallise. Within the honey, it is glucose which forms crystals, fructose remains in solution. For granulation to occur, it is essential that the molecules within the solution are able to move about freely, so enabling them to form crystals. There are a number of factors which influence the rate and type of granulation:

- *temperature*. Normally there is no granulation above 30 °C and below 10 °C granulation slows down, stopping altogether below 4.5 °C. The high temperature keeps the glucose in solution, the low temperature slows down movement of molecules within the honey, preventing the formation of crystals.
- *glucose/water ratio*. The higher it is the more rapid will be the granulation.

- *glucose/fructose ratio*. Honeys with a fructose content equal to, or higher than, the glucose content are slow to granulate. An example is the honey from false acacia (*Robinia pseudoacacia*). Conversely, honeys with a high glucose content, eg, oilseed rape (*Brassica napus*) granulate very rapidly.
- *the viscosity of the honey*. Where the honey is very viscous, granulation is slow. This is because the movement of the molecules within the honey is slowed.
- *presence of nuclei around which crystals form*. These can be specks of dust, clumps of pollen grains or very small crystals.
- *stirring honey* will speed the rate of granulation as it will distribute the nuclei evenly.
- *the speed of granulation* influences the size of the crystals produced, so slow granulation leads to large coarse crystals and rapid granulation produces small crystals and a smoothly granulated honey.

Practical application to honey processing

Honeys may be kept liquid by storing in very warm, or cold, conditions. Exposing honey to high temperatures will damage it in other ways, as we have seen above. Cold storage is expensive. So, the best that can usually be done is to store honey in a cool place. It is a pity that cellars went out of fashion in the UK, because they could often provide very good storage conditions.

To produce a 'good' granulation, ie, a smooth, non-gritty one, with small crystals, honey should be kept at 14 °C, the average temperature at which granulation is fastest, should be stirred frequently and, if necessary, should be 'seeded' by adding a quantity (usually about 10% by weight) of a honey with small crystals, which has been softened so that it can be incorporated into the liquid bulk. The small crystals will act as nuclei which will induce the formation of small crystals within the whole quantity of honey. Frequent stirring will ensure that they are well distributed and will break up any clumps of crystals which may form.

Granulated honey can be returned to its former, liquid state by heating, but care must always be taken to ensure that overheating does not occur. In commercial honey production, honey can be heated and cooled rapidly but that is not so easy for the small-scale beekeeper.

It is very difficult to generalise about honey granulation because every honey sample is different, but any beekeeper will soon become familiar with 'their' honeys and will be able to process them accordingly.

FERMENTATION – FRIEND AND FOE

The chemical process of fermentation can be defined as the enzymatic and anaerobic decomposition of organic substances, other than proteins, usually by bacteria or yeasts to yield simpler organic compounds. The end-products can be various. However, generally, it is accepted as being the action of yeasts on sugar to give ethanol (ethyl alcohol) and carbon dioxide. This, strictly speaking, is alcoholic fermentation.

Useful chemistry

Fermentation has been used by man for a long time. Because one of the end-products is alcohol, it is used to produce wines and beers. Of course, it depends on your point of view whether this is good or bad! It is likely that man's oldest intoxicating liquor was mead, produced by the fermentation of honey. The process is also used in bread-making where the carbon dioxide produced is used to make the bread 'rise', that is, increase in volume with the characteristic spongy texture associated with bread.

Honey spoilage

Apart from mead-making, the main interest in fermentation, from the point of view of the beekeeper, is as a process which can spoil stored honey and it is this aspect which concerns us here.

Conditions necessary for fermentation of stored honey

These can be listed quite simply but we need to expand on them:

- the presence of osmophilic yeasts in sufficient quantity in the honey
- adequate warmth for yeast growth and reproduction
- a sufficiently high water content of the honey to allow the yeasts to grow.

Osmophilic yeasts

Yeasts are simple one-celled micro-organisms which multiply by budding. This means that an outgrowth from the cell grows until it is constricted and becomes a separate cell. It may remain attached to the 'parent' cell and the same process may occur many times, the resultant mass consisting of many cells which, although still attached to one another, are each quite independent. All this is interesting but fairly irrelevant, I hear you saying, but it helps to know what is going on. There are many different kinds of yeast and those adapted for living in very concentrated solutions, such as honey, are called *osmophilic* yeasts. The yeasts found in honey are found everywhere where there is honey, such as the bodies of bees, the equipment and structure of the honey house, the soil in the apiary and so on. They are also present in nectar. It is impossible to keep them out of honey and they are always present to a greater or lesser degree. The only recognised way of removing them is by heating the honey to a high enough temperature to kill them and then ensuring that the honey does not become recontaminated. In commercial practice this may be done by 'pasteurizing' the honey, usually to 71 ° C instantaneously and then cooling it rapidly.

Temperature

Like most living organisms, yeasts require an equitable temperature to grow and multiply successfully. The range at which they are most active is 18–21 °C but there will be some activity below this.

Moisture content

Broadly speaking, honey with a moisture content of less than 17% will not ferment and honey with a moisture content of more than 19% will almost certainly ferment, given time and other satisfactory conditions.

Understanding why some honey ferments

I have listed the conditions which are necessary for fermentation but it is important to understand the interaction between them. If, for example, there is a very low level of yeast in a honey, even at a moisture content of 19% it may be safe. If the moisture content is relatively high, there is plenty of yeast present, but the honey is kept at a temperature below 13 °C, it will not ferment.

Anyone who sells honey must be aware of current legislation

Effect of granulation

When the glucose in honey begins to crystallize, some of the water is used within the crystal structure. Unfortunately there is a larger amount of water left behind in the liquid part of the honey between the crystals. This means that the liquid part of the honey will have a higher moisture content than the totally liquid honey had and this can be sufficient to allow fermentation to start. So, granulated honey, and particularly partially-granulated honey, is more likely to ferment than liquid honey. The greatest danger period for the small-scale beekeeper is the spring when the temperature in the storage area begins to rise and the honey has all granulated. The buckets of stored honey then begin to bubble and the honey gives off a strange aroma.

Prevention of fermentation

There are three main avenues open to the beekeeper:

- controlling the yeasts
- avoiding the ideal temperature for yeast growth
- keeping the moisture content below 17%.

All honeys will contain yeasts, as we have seen. Application of heat can kill them and the pasteurisation process, mentioned above, does that. For the small-scale beekeeper, heating liquid honey to 60 °C for one hour, with rapid cooling, will be effective (and also destroy tiny crystals which will start granulation). It is important to ensure that there is no more contamination with yeasts after this process.

We should strive to store our honey in cool conditions, below the threshold which allows the yeasts to grow rapidly. If granulated honey is needed, then it should be stored at the optimum temperature for granulation then, once completely solid, stored at a lower temperature. All of this is obviously very difficult for the small-scale beekeeper and too expensive for the commercial man.

Moisture content is really the key to controlling fermentation. First, ensure that the moisture content of the honey you are extracting is 17% or lower. By and large, if you only extract capped honey, it will be safe, but in these days of oilseed rape production that is a counsel of perfection and we cannot avoid extracting some unsealed honey. This will be fine, in most cases, if tested by the 'shaking' method, but it is much more reliable to use a hydrometer or refractometer. I know they cost money, but how many buckets of fermented honey will you need to pay for such an item? Having extracted honey with a low moisture content, it must be stored in full, sealed containers so that damp air cannot gain access to it, otherwise it will absorb moisture and all the care taken when extracting will be wasted.

What to do with fermenting honey

This advice is similar to finding an item on collecting swarms at the end of instructions on swarm control, but, despite all the care taken with honey, sooner or later most beekeepers find they have some fermenting honey. What do you do with it? If the fermentation is slight, affecting only the top of the honey, the worst can be skimmed off and the honey fed back to the bees. They will reprocess it for you. But do not feed it as winter stores. If it is really horrible and the bulk of the honey is affected, it makes a good compost accelerator. Cover it well so that the bees cannot gain access to it. Then go out and buy a refractometer!

Of course, the other product of fermentation is mead and this can be enjoyed, especially in the cold winter months when it will revive memories of long summer days and beekeeping, but remember that too much enjoyment of this hive product may leave you with the belief that fermentation is indeed more foe than friend.

LEGAL REQUIREMENTS FOR THE SALE OF HONEY

When most of us start keeping bees, nothing is further from our minds than legal responsibilities but, once our colony numbers increase, producing a few buckets of honey each year, the urge, and indeed the necessity, to sell it develops. It seems easy enough. Many of our friends will happily buy it from us and there are some old jam jars stored in boxes in the shed. The local shop hears about this wonderful new source of honey and away you go – but beware – there are numerous pitfalls for the unwary and you may find a Trading Standards Inspector or Environmental Health official standing on your doorstep because we must always remember that honey is a food. The rules and regulations applying to honey are common sense, for the most part, and for the protection of customers and beekeepers alike. They fall into broad categories:

- weights and measures
- hygiene
- labelling.

CHAPTER 13

The main pieces of legislation in the UK are as follows.

- *The Food Safety Act (1990)* and some of its Regulations.
- *The Honey (England) Regulations 2003* (amended in 2005 and 2011). (If you live in Northern Ireland, Scotland or Wales you have your own.) These Regulations are essential reading for anyone who sells honey. They are made under the Food Safety Act 1990, implement the European Council Directive 2001/110/EC and replace the 1976 Regulations, with minor changes.
- *The Food Labelling Regulations 1996* address the labelling of various types of food.
- *The Food (Lot Marking) Regulations 1996* which require each batch of honey to be given a lot number so that it can be traced back to its source.
- *The Weights and Measures Act 1985* was amended in 1994 by various Orders and Regulations. These lay down the regulations on weight but, put simply, states that you must not sell underweight.

This is not an exhaustive list and is very simplified but there are a few general points that I want to make.

Honey must be extracted, processed and handled in hygienic conditions. There are strict rules included in the Food Safety Act.

If you are selling your honey, it must meet composition requirements including maximum water content, maximum sucrose content, maximum HMF and minimum diastase content. These are the ones over which you have some control, even though you cannot measure them. Others are dependent upon the nature of the honey as produced by the bees, but must still be observed.

Labelling is simpler to understand and the one most likely to affect the small producer, but there are still pitfalls for the unwary. The following must appear on the label:

- The word 'Honey', but this can be qualified and there are other specified descriptions set out such as 'comb honey'. Always be sure, if you use an adjective to describe your honey, that it is accurate. If you decide to embellish your label with pictures of flowers they must be relevant to the contents of the jar and must not mislead, eg, a picture of a heather moor on honey which is not heather.
- A name and address, either your own, or a packer.
- The weight in metric units. This lettering has to be a certain size, dependent on the size of the pack.
- Lot number. The purpose of the lot number is to allow any

batch of honey with which there are problems to be traced and recalled, so you must keep records and it pays to keep the batches small. The number must be preceded by the letter 'L'.
- 'Best before' date. This is left to your discretion. Do you give a long date or a shorter one designed to focus the mind of the purchaser and ensure that they eat it up and come back for more?
- Country of origin. Assuming that your bees are in the UK, it will state 'Product of UK'.

It is important that all the above information is indelible and is within the field of view of the purchaser.

This is a very brief indication of the legal requirements in the UK. I am not a lawyer and regulations change. Trading Standards departments are always helpful and can offer advice, as can the beekeeping organisations. Other sources of information are the Internet and various leaflets. Other countries have their own legislation and it is always the responsibility of the seller to abide by the rules. Ignorance is never a defence.

GLOSSARY

In this glossary I have attempted to give guidance on pronunciation of some words by putting phonetic spelling in brackets. The accented syllable is in *italics*. There are often alternative ways of pronouncing words. Note that some of the definitions apply to the subject matter of this book and that words may have a wider, or different, meaning in other branches of biology.

acaricide (*a*-ka-ri-side)
A substance which will kill mites.

acarine (*a*-ka-rin)
Disease of adult honey bees caused by *Acarapis woodi*.

Aculeata (a-kool-ee-*ai*-ta)
Division of the Apocrita having the ovipositor modified into a sting.

alkaloid (*alk*-al-oyd)
Group of basic nitrogenous organic compounds produced by plants.

ambophily (am-*boff*-ill-ee)
Pollination by both air currents and insects.

amino acid (a-*meen*-oh acid)
Organic compound containing the amino group (-NH$_2$) and from which proteins are built.

amoeba
Genus of the protozoa, many free-living but some parasitic.

androecium (and-*ree*-see-um)
Collective term for the stamens of a flower, cf. gynaecium.

anemophily (an-ee-*moff*-ill-ee)
Pollination by air currents (wind).

Angiospermae (an-jee-oh-*sperm*-ee)
Group of plants whose seed is borne in a vessel, ie, the flowering plants.

anther
Terminal part of a stamen, containing pollen.

Apocrita (ap-*ok*-ri-ta)
Sub-order of the Hymenoptera having a petiole between the propodaeum and the second abdominal segment.

apomixis (ay-poh-*mix*-is)
Usually in plants, reproduction which appears to be sexual but occurs without fertilisation.

Arachnida (a-*rack*-nid-a)
Class of the Arthropoda including the spiders, mites, ticks, etc.

Arthropoda
Phylum in the animal kingdom, including the Classes Insecta, Arachnida, Crustacea, etc.

auricle (o-*rick*-ul)
Flattened area, on the upper surface of the basitarsus, which collects the pollen combed from the pollen brushes by the rastellum.

bacterium
Huge group of unicellular organisms lacking a nuclear membrane.

basitarsus (bay-zi-*tar*-sus).
Proximal section of the tarsus.

calyx
Collective name for the sepals of a flower.

Class
Large section of a phylum, eg, Insecta.

corbicula (kor-*bik*-yule-a)
Pollen basket of the worker honey bee, situated on the outside of the hind tibia and used for transporting pollen loads.

corolla
Collective name for the petals of a flower.

cross breeding
Mating between unrelated individuals to produce genetically variable offspring, cf. line breeding.

cross-pollination
Receipt of pollen from a different plant of the same species, cf. self-pollination.

cuticle
Outer covering of the larva, pupa and adult honey bee.

deoxyribonucleic acid (DNA)
Genetic material of most living organisms. Major component of chromosomes.

GLOSSARY

diastase
Enzyme breaking down starch; also known as amylase.

dioecious (die-*ee*-shus)
Having separate male and female flowers on different plants, cf. monoecious.

Diptera
Insect Order of two-winged flies.

dysentery
Condition where honey bees void faeces in and on the hive.

ectoparasite
Parasite which lives on the outside of its host.

endoparasite
Parasite which lives internally in its host.

entomophily (ent-oh-*moff*-ill-ee)
Transfer of pollen by insects.

epithelium (eppy-*theel*-ee-um)
Tissue which covers a structure, usually made of closely packed cells.

eukaryote (you-*ka*-ree-oht)
Organism consisting of cells where the genetic material is contained in a distinct nucleus, cf. prokaryote.

eusocial
Of a colony of insects, consisting of an egg-laying queen and two or more generations of adult females which function as workers.

exine
Protective outer covering of a pollen grain, cf. intine.

extra-floral nectary
Nectar-producing structure found on a part of a plant other than a flower.

Family
In classification, group of related organisms, between Order and Genus.

fat bodies
A diffuse tissue found throughout the body of the larva and adult honey bee as a food store.

fertilisation
Fusion of a male gamete and a female gamete to produce a zygote.

filament
Part of a stamen, the stalk which carries the anther.

flagellum (fla-*jell*-um)
Extension of a cell, often used in locomotion.

fuchsin
Red stain used in microscopy.

fumagillin (fume-a-*gill*-in)
Antibiotic obtained from *Aspergillus fumigatus* and effective in suppressing Nosema. <u>It is no longer available</u>.

fungus
Eukaryotic organism, lacking chlorophyll and, typically, composed of hyphae.

gamete
Reproductive cell.

Genus
In classification, a group of closely related organisms between Family and Species.

glossa
Central, tubular structure of the proboscis.

granulation
Formation of crystals in a liquid solution.

graticule
Scale used for making measurements through the microscope.

gynaecium (gigh-*nees*-ee-um)
Collective term for the female parts of a flower, cf. androecium.

haemolymph (*hee*-moh-lymf)
Blood of the honey bee.

haplotype (*hap*-loh-type)
Within a species, the range of different gametes.

heterostyly (het-err-oh-*stigh*-lee)
Having two, or more, forms of flower on different plants with respect to styles and anthers and, sometimes, pollen.

hydrocarbon
Organic compound containing only the elements carbon and hydrogen.

hydrometer
Instrument used to measure density or relative density of liquids.

hydroxymethylfurfural (HMF)
Breakdown product of fructose, its production being increased by storage and high temperatures.

Hymenoptera (high-men-*op*-te-ra)
Order containing bees, wasps, ants and allied insects with two pairs of membranous wings.

hypha (*high*-fa)
Of a fungus, a tubular thread.

hypopharyngeal gland (high-poh-fa-rin-*jee*-al gland)
One of a pair of glands in the head of the worker honey bee.

inquiline
Organism sharing the nest of an unrelated species without apparent harm to either.

Integrated Pest Management (IPM)
System of control of a condition/disease where a range of different methods is employed at different stages.

intine
Inner coat of a pollen grain, cf. exine.

GLOSSARY

intracellular
Occurring within a cell.

Isle of Wight disease
Condition causing the deaths of many colonies of honey bees during the early part of the twentieth century.

Kingdom
In classification, the largest group of organisms. Most systems recognise five kingdoms covering all living organisms on the planet.

Lepidoptera
Insect Order including the butterflies and moths.

line breeding
Mating of closely related individuals, leading to consistency among the offspring, cf. cross breeding.

lipoprotein (lip-oh-*proh*-teen)
Substance which is a complex of fat and protein.

Malpighian tubule (mal-*pig*-ee-an tubule)
One of the long, narrow excretory structures of the honey bee.

melissopalynology (mel-iss-oh-pal-in-*o*-loj-ee)
Study of pollen content of honeys.

mesothorax (*mee*-soh-thor-racks)
Second thoracic segment.

microlepidoptera
Large group of very small moths within the Order Lepidoptera.

Microsporidia (mike-roh-spo-*rid*-ee-a)
Group of intracellular parasites, affecting the intestinal systems of insects and other animals.

mite
Member of the Order Acari, within the Arachnida. Important group including many pathogens and vectors of disease.

monoecious (mon-*ee*-shus)
Having separate male and female flowers on the same plant, cf. dioecious.

monofloral honey
Honey derived from a single type of flower.

mycelium (migh-*seel*-ee-um)
Collective term for a mass of fungal hyphae.

nectary
Gland in plants secreting a sugary liquid. Usually associated with flowers.

neonicotinoid (knee-oh-ni-*cott*-in-oyd)
One of a group of chemical insecticides, related to nicotine and acting systemically in the plant.

Nosema
Disease of adult bees affecting the ventriculus.

notum
Alternative, and usual, name for a tergum in a thoracic segment.

nymph
Larval stage in the development of an insect, or other Arthropod, which resembles the adult but is sexually immature.

obligate parasite
Parasite which cannot survive independently of its host.

oenocyte (*een*-oh-sight)
Cell found in the fat body and elsewhere, particularly the wax glands. Involved in production of lipoproteins.

osmophilic (oz-moh-*fill*-ik)
Able to withstand high osmotic pressure, eg, yeasts which grow in honey.

ovary
a) In flowers the part of the gynaecium producing, and housing, the ovules.
b) In female honey bees, the part of the female reproductive system producing the eggs.

ovule
Structure within the ovary developing into a seed after fertilisation.

oxytetracycline (OTC)
Antibiotic used to treat European foul brood.

paralysis
Number of different viral conditions resulting in a variety of symptoms, including an inability to move properly.

parasite
Organism living in or on another organism (its host) and obtaining nourishment at the expense of the host.

pathogen
Disease-causing organism.

peritreme (*perry*-tream)
A tube enabling the female varroa mite to breathe while submerged in brood food. Can be likened to a snorkel.

peritrophic membrane (peri-*trofe*-ick membrane)
Gelatinous envelope surrounding the contents in the ventriculus.

petal
Individual part of the corolla of a flower.

petiole
a) Narrow constriction between the propodaeum and the second abdominal segment in the Apocrita.
b) Leaf stalk.

GLOSSARY

pH
Logarithmic scale for expressing acidity or alkalinity of a solution.

phloem (*flow*-em)
Tubes within a plant carrying plant sap.

phoretic stage
Of a life cycle, the migratory stage.

Phylum
In classification, large group between Kingdom and Class.

pollen
Structure produced by the anther and containing the male gamete.

pollen brush
Rows of bristles on the inside of the basitarsus of the hind leg of a worker bee, used in pollen collection.

pollen press
Collection of structures between the tibia and basitarsus of a worker bee, used to push pollen into the corbicula.

pollination
Movement of pollen from an anther to a stigma in a plant of the same species.

prepupa
Short stage in honey bee's life cycle between larval and pupal stages; the period during which the fifth moult occurs.

prokaryote (proh-*karr*-ee-oht)
Large group of unicellular organisms whose DNA is not contained in a nuclear membrane. Includes the Bacteria, cf. eukaryote.

propolis
Bee 'glue', derived from resinous exudates of plants by worker honey bees.

protandry
Condition in flowers where the anthers ripen before the stigmas, cf. protogyny.

prothorax
First thoracic segment.

protogyny (proh-toh-*gigh*-nee)
Condition in flowers where the stigmas ripen before the anthers, cf protandry.

protozoa
Large group of unicellular or colonial animals, mostly living in damp or aquatic environments, but some parasitic on other animals.

pupa
Stage, in the life cycle of insects showing complete metamorphosis, between larva and imago.

rastellum
Fringe of downward-pointing, stiff bristles on the hind tibia of the worker bee, used to scrape pollen from the pollen brush onto the auricle, also known as rake.

receptacle
Part of the flower to which all the floral parts are attached, also known as torus.

refractometer
Instrument used to determine water content of honey by measuring its refractive index.

safranin
Red stain used in microscopy.

scopa
Pollen-collecting structure on either the legs or beneath the abdomen of solitary bees.

self-pollination
Pollination with pollen from the same plant, usually the same flower, cf. cross-pollination.

sepal
Individual part of the calyx of a flower.

Species
In classification, an interbreeding group of organisms.

spiracle (*spi*-rack-ul)
Opening in the body wall of an insect, allowing gases to flow in and out of the respiratory system.

stamen
Individual part of the androecium of a flower, comprising an anther and filament.

Standstill Order
Legal order prohibiting the movement of hives, bees and other equipment from an apiary with a notifiable disease.

stigma
In a flower, the part of the gynaecium which receives the pollen.

style
In a flower, the connection between the ovary and the stigma, through which the pollen tube grows.

synergistic (sin-err-*jist*-ick)
The increased effect of two substances working together.

thelytoky (thel-ee-*toke*-ee)
Parthenogenetic production of females.

trachea (track-ee-a)
Tube forming part of the respiratory system and carrying gases.

trophallaxis (trohf-al-*aks*-is)
Food sharing between individuals in a colony of honey bees.

***Tropilaelaps* spp.** (troppy-*lie*-laps species)
A genus of parasitic Asiatic mites, originally infesting *Apis dorsata*.

GLOSSARY

vector
Organism housing parasites and transmitting them from one host to another.

ventriculus (vent-*rick*-you-lus)
True stomach, where digestion takes place.

virus
One of a group of minute obligate parasites only able to multiply inside a cell of their host. They contain DNA or RNA, rarely both.

yeast
Group of unicellular fungi occurring everywhere, including in honey, and capable of breaking down glucose to alcohol and carbon dioxide, a process known as fermentation.

REFERENCES AND FURTHER READING

Adam, Brother (1983) *In Search of the Best Strains of Bees*. Northern Bee Books, Mytholmroyd.

Aston, D and Bucknall, S (2004) *Plants and Honey Bees: their relationships*. Northern Bee Books, Mytholmroyd.

Bailey, L (1971) *Honey Bee Paralysis; Retrospect and Prospect*. Central Association of Beekeepers.

Bailey, L & Ball, BV (1991) *Honey Bee Pathology*. 2nd edn. Academic Press, London.

Ball, B (1995) *Varroa jacobsoni: host-parasite-pathogen interactions*. Central Association of Beekeepers.

Benton, Ted (2006) *Bumblebees: the natural history & identification of the species found in Britain*. Collins, London.

Blackman, R (1974) *Aphids*. Ginn, London.

Chapman, RF (2013) *The Insects: structure and function*. 5th edn. Cambridge University Press, New York.

Crane, E (Ed) (1975) *Honey: a comprehensive survey*. Heinemann, London.

Daintith, J & Martin, E (1999, 2003 printing) *A Dictionary of Science*. Oxford University Press, Oxford.

Defra (2005) *Managing Varroa*. Department for Environment, Food and Rural Affairs, London.

Else, GR (1994) 'Identification: Social Wasps'. *British Wildlife*, **5**: 304–311.

Frisch, Karl von (1950, 1983 printing) *Bees: their vision, chemical senses and language*. Jonathan Cape Ltd, London.

Harley, Madeline M (2004) *Pollen – all packed up and ready to go*. Central Association of Beekeepers.

Hodges, D (1984) *The Pollen Loads of the Honeybee: a guide to their identification by colour and form*. IBRA, London.

Hooper, T and Taylor, M (1988) *The Beekeeper's Garden*. Alphabooks Ltd.

Kirk, WDJ & Howes, FN (2012) *Plants and Beekeeping: a guide to the plants that benefit the bees of the British Isles*. IBRA, Cardiff.

REFERENCES AND FURTHER READING

Kirk, William DJ (1994) *A Colour Guide to Pollen Loads of the Honey Bee*. IBRA, Cardiff.

Knight, Albert (2005) *A Rational Approach to the Honey Bees of Britain*. Central Association of Beekeepers, Poole.

MAFF (1996) *Foul Brood Disease of Honey Bees: Recognition and Control*. Ministry of Agriculture, Fisheries and Food.

MAFF (1999) *Statutory Procedures for Controlling Foul Brood*. Ministry of Agriculture, Fisheries and Food.

Martin, Stephen J (2003) *Man-made Beekeeping Problems: is the 'Capensis problem' a good case in point?* Central Association of Beekeepers, Poole.

Morse, Roger A and Flottum K, eds (1997) *Honey Bee Pests, Predators, & Diseases*. AI Root Company, Medina, OH.

O'Toole, C and Raw, A (1991) *Bees of the World*. Blandford, London.

Proctor, M and Yeo, P (1973) *The Pollination of Flowers*. Collins, London.

Ribbands, R (1953) *The Behaviour and Social Life of Honeybees*. Bee Research Association Ltd, London.

Seeley, Thomas D (1995) *The Wisdom of the Hive: the social physiology of honey bee colonies*. Harvard University Press, Cambridge, MA, USA.

Sawyer, Rex (1981) *Pollen Identification for Beekeepers*. University College Cardiff Press, Cardiff.

Taber, Steve (1987) *Breeding Super Bees*. AI Root Company, Medina, OH (extensive bibliography).

Tautz, J (2008) *The Buzz about Bees: biology of a superorganism*. Springer, Berlin, Germany.

Wilson, Edward O (1971) *The Insect Societies*. The Belknap Press of Harvard University Press, Cambridge, MA.

Winston, Mark (1987) *The Biology of the Honey Bee*. Harvard University Press, Cambridge, MA.

Yeo, PF & Corbet, SA (1983) *Solitary Wasps*. Cambridge University Press, Cambridge, UK.

Microscopes, equipment, solvents, etc, are obtainable from
Brunel Microscopes Ltd
Tel: 01249 462655; e-mail: mail@brunelmicroscopes.co.uk; website: www.brunelmicroscopes.co.uk
They are helpful and patient.

INDEX

Acarapis woodi, 60, 70–1,
　see also acarine
acarine, 18, 19, 21, 60, 71–3
acetic acid, *see* ethanoic acid
Achroia grisella, *see* moths, wax
Aculeata, 4, 5
Aethina tumida, 91
AFB, *see under* foul brood
amino acids, 36
amoeba, 60, 63, 68–70
Angiosperms, 11–12, 111,
　135–6
antibiotics, 40, 51, 54
ants, 4
Apidae, 4, 5, 9
Apis mellifera capensis, 21–2
A. m. carnica, 18
A. m. caucasica, 18–19
A. m. ligustica, 19
A. m. mellifera, 20–1
A. m. scutellata, 21
Apis, 4, 5
Apocrita, 3–4, 5
Apoidea, 4, 5
Arthropoda, 2–3, 5, 42, 70
Ascosphaera apis, *see* chalkbrood
Aspergillus spp., *see* stonebrood

Bacillus thuringiensis, 95
bacteria, 40, 47–8, 51–4, 95
　beneficial, 34, 40
　see also under foul brood
Bailey comb change, 68
beans, field, 106
'bee bread', 107, 156
bees,
　Africanised, 21–3
　scout, 154
　solitary, 8–9, 12–13
　subspecies of, 15–22
　winter, 65, 110
beeswax, 144–9
　chemistry of, 145–8
　production of, 148–9
　properties of, 145
borage, 106
Braula coeca, 89–90
breeding, 23–31, 87
　cross, 28–9, 30
　line, 28, 30
brood, 159
　bald, 94
　sealed, 38, 48
　unsealed, 37
　see also larvae
'Buckfast' bees, 22–3
bumblebees, 9–11
　cuckoo, 10

chalkbrood, 55–7
chemicals, agricultural, 106–9
classification, 1–5

Colony Collapse Disorder (CCD),
　62, 100
comb, replacing, 68, 70, 82

dances, recruitment, 152–3
density, 164
diastase, 163
disease, control of, 46–7, 50–1,
　54, 57, 68, 70, 73
　monitoring for, 37
　notifiable, 50–1, 54
　spread of, 49, 53, 54, 65
drifting, 19, 49, 54, 76
drones, 27, 29
　diploid, 30
dysentery, 65, 67

EFB, *see under* foul brood
enzymes, 161–3
ethanoic acid, 35, 55, 94–5
evolution, of bees, 11–14
　of flowering plants, 11–12,
　117
fermentation, 159–60
fertilisation, 113–14, 117
flowers, structure of, 111–15,
　118–22
forage for bees, 103–6
foraging behaviour, 152–6
foul brood, American (AFB)
　47–51

183

foul brood, European (EFB), 51–5, 103
fungi, 41, 55–7, *see also* yeast
fungicides, 107

Galleria mellonella, *see* moths, wax
galls, 4
gaster, 3–4
glands, hypopharyngeal, 46, 65, 161–2

hive beetle, small, *see Aethina tumida*
Honey (England) Regulations 2003, 159, 172–3
honey, 91, 120, 124–5, 159–73
    comb, 90, 94
    composition of, 162–4, 166–7
    fermentation of, 168–71
    granulation of, 166–8, 170
    hygroscopicity of, 165
    identification of, 136
    laws relating to sale of, 171–3
    physical properties, 164–70
    poisonous, 129–30
    unpleasant, 129
honeydew, 131–3
hornet, 7–8, 96
    Asian, 11, 96–7
hybrids, 15, 22, 24, 29, 30
hydrometer, 164, 170
hydroxymethylfurfural (HMF), 163
hygiene, 35, 172
Hymenoptera, 3, 5

immunity, 34, 36, 67, 81, 101
Insecta, 3, 5, 42–3
insecticides, 107–8
insemination, instrumental, 29–30
Integrated Pest Management (IPM), 84–7
Isle of Wight disease, 19, 60, 70

juvenile hormone, 151
labelling, 172–3
larvae, 37–8, 45–6, 47, 51–7
laws relating to beekeeping, 50–1, 54, 88, 91, 171–3
Linnaeus, Carolus, 1
louse, bee, *see Braula coeca*

Malpighamoeba mellificae, *see* Amoeba
mating, 27, 29
Melissococcus plutonius, *see* foul brood, European
melissopalynology, 136
mice, 34, 97–8
microscopes, 40, 78, 136
microsporidia, 41–2, 63–8
mites, 42–3, 70–3, 75–88
moths, wax, 34, 91–5
    biological control of, 95

nectar, 8, 9, 10, 11, 122, 123–31, 152
    change to honey of, 160–2
    factors affecting secretion of, 126–8
    production of, 113, 118, 119, 120
    sugars in, 161–2
    transport of, 154–5
nectaries, 118, 119, 120, 125–6
    extra-floral, 130–1
Nosema apis, 63, 64–6, 69, *see also* nosema
Nosema ceranae, 36, 64, 66–7, *see also* nosema
nosema, 18, 19, 60, 64–8, 102

Paenibacillus larvae, *see* foul brood, American
parasitic mite syndrome, 81, 103
Parasitica, 4
pathogens, 33–5, 36
    movement of, 50
    see also individual diseases

pesticides, 36
pests, 91–100
    notifiable, 88, 91
petiole, 3
plants, flowering, *see* Angiosperms
pollen, 9, 10, 11, 50, 67, 93, 112, 114, 135–41
    collection of, 155
    composition of, 140
    identification of, 136–7, 139–40
    microscope slides of, 137–9
    poor supply of, 36, 68
    structure of, 135, **139**
pollination, 9, 113–17, 119, 122, 140
propodaeum, 3
propolis, 19, 25, 34, 141–4, 152, 156–7
    collection of, 157
    composition of, 141–2
    human uses for, 143
    use by bees of, 142, 143–4
Protozoa, 42, 68–70

queen, 25–7, 29–31, 85, 90

rape, oilseed, 105, 106
records, 25
refraction, 165–6
refractometer, 166, 170–1
relative density, 164
robbing, 19, 35, 49, 54, 76

sacbrood, 45–7
sociality, evolution of, 12–14
sting, 4
stonebrood, 57
stress, 35–6, 55, 67–8, 82, 109
subspecies of *Apis mellifera*, 15–23
swarming, 24–5
swarms, 49, 50, 54, 76
    shook, 54
Symphyta, 4

INDEX

taxon, 1
termites, 14
thixotropy, 120, 165
Tropilaelaps spp., 87–8

uniting, 50

Varroa destructor, see varroa
varroa, 36, 37, 40, 75–87, 94, 101
   and acarine, 73
   and viruses, 59, 61–3, 102
   chemical treatments for, 81–4, 107
   life cycle of, 76–8
   monitoring for, 84, 86
   resistance to chemicals of, 82–4
Vespa velutina, 11, 96–7
virus, acute bee paralysis, 59, 62
   black queen cell, 47, 67
   chronic bee paralysis (CBPV), 59–61
   deformed wing (DWV), 59, 61–2
   slow bee paralysis (SBPV), 59, 61
viruses, 39–40, 45–7, 59–63, 67, 69, 80, 102
viscosity, 164–5, 167
vitellogenin, 151

wasps, 4, 34, 95–6
   gall, 4
   social, 7–8
   solitary, 6–7
water, collection and use of, 144, 152
wax, *see* beeswax
woodpeckers, 98–9

yeast, 168–70